ACOUSTICAL AND ENVIRONMENTAL ROBUSTNESS IN AUTOMATIC SPEECH RECOGNITION

THE KLUWER INTERNATIONAL SERIES IN ENGINEERING AND COMPUTER SCIENCE

VLSI, COMPUTER ARCHITECTURE AND DIGITAL SIGNAL PROCESSING

Consulting Editor

Jonathan Allen

Latest Titles

The Design of Communicating Systems: A System Engineering Approach,
C. J. Koomen
ISBN: 0-7923-9203-5
Parallel Algorithms and Architectures for DSP Applications,
M. A. Bayoumi, editor
ISBN: 0-7923-9209-4
Digital Speech Processing: Speech Coding, Synthesis and Recognition
A. Nejat Ince, editor
ISBN: 0-7923-9220-5
Sequential Logic Synthesis, P. Ashar, S. Devadas, A. R. Newton
ISBN: 0-7923-9187-X
Sequential Logic Testing and Verification, A. Ghosh, S. Devadas, A. R. Newton
ISBN: 0-7923-9188-8
Introduction to the Design of Transconductor-Capacitor Filters,
J. E. Kardontchik
ISBN: 0-7923-9195-0
The Synthesis Approach to Digital System Design, P. Michel, U. Lauther, P. Duzy
ISBN: 0-7923-9199-3
Fault Covering Problems in Reconfigurable VLSI Systems, R.Libeskind-Hadas, N. Hassan, J. Cong, P. McKinley, C. L. Liu
ISBN: 0-7923-9231-0
High Level Synthesis of ASICs Under Timing and Synchronization Constraints
D.C. Ku, G. De Micheli
ISBN: 0-7923-9244-2
The SECD Microprocessor, A Verification Case Study, B.T. Graham
ISBN: 0-7923-9245-0
Field-Programmable Gate Arrays, S.D. Brown, R. J. Francis, J. Rose, Z.G. Vranesic
ISBN: 0-7923-9248-5
Anatomy of A Silicon Compiler, R.W. Brodersen
ISBN: 0-7923-9249-3
Electronic CAD Frameworks, T.J. Barnes, D. Harrison, A.R. Newton, R.L. Spickelmier
ISBN: 0-7923-9252-3
VHDL for Simulation, Synthesis and Formal Proofs of Hardware, J. Mermet
ISBN: 0-7923-9253-1
Wavelet Theory and its Applications, R. K. Young
ISBN: 0-7923-9271-X
Digital BiCMOS Integrated Circuit Design, S.H.K. Embabi, A. Bellaouar, M.I Elmasry
ISBN: 0-7923-9276-0
Design Automation for Timing-Driven Layout Synthesis, S. S. Sapatnekar, S. Kang
ISBN: 0-7923-9281-7

ACOUSTICAL AND ENVIRONMENTAL ROBUSTNESS IN AUTOMATIC SPEECH RECOGNITION

by

Alejandro Acero
Carnegie Mellon University

foreword by

Richard M. Stern

KLUWER ACADEMIC PUBLISHERS
Boston/Dordrecht/London

Distributors for North America:
Kluwer Academic Publishers
101 Philip Drive
Assinippi Park
Norwell, Massachusetts 02061 USA

Distributors for all other countries:
Kluwer Academic Publishers Group
Distribution Centre
Post Office Box 322
3300 AH Dordrecht, THE NETHERLANDS

Library of Congress Cataloging-in-Publication Data

Acero, Alejandro.
Acoustical and environmental robustness in automatic speech recognition / by Alejandro Acero.
p. cm. -- (The Kluwer international series in engineering and computer science. VLSI, computer architecture, and digital screen processing)
Includes bibliographical references and index.
ISBN 0-7923-
1. Automatic s
I. Title. II. Serie
TK7882 . S65A24
621 . 39 ' 9--dc20 -31211
CIP

Printed on acid-free paper.

Printed in the United States of America

Table of Contents

List of Figures

List of Tables

Foreword

The need for automatic speech recognition systems to be robust with respect to changes in their acoustical environment has become more widely appreciated in recent years, as more systems are finding their way into practical applications. Although the issue of environmental robustness has received only a small fraction of the attention devoted to speaker independence, even speech recognition systems that are designed to be speaker independent frequently perform very poorly when they are tested using a different type of microphone or acoustical environment from the one with which they were trained. The use of microphones other than a "close-talking" headset also tends to severely degrade speech recognition performance. Even in relatively quiet office environments, speech is degraded by additive noise from fans, slamming doors, and other conversations, as well as by the effects of unknown linear filtering arising reverberation from surface reflections in a room, or spectral shaping by microphones or the vocal tracts of individual speakers. Speech-recognition systems designed for long-distance telephone lines, or applications deployed in more adverse acoustical environments such as motor vehicles, factory floors, or outdoors demand far greater degrees of environmental robustness.

There are several different ways of building acoustical robustness into speech recognition systems. Arrays of microphones can be used to develop a directionally-sensitive system that resists interference from competing talkers and other noise sources that are spatially separated from the source of the desired speech signal. The use of peripheral signal processing schemes based on computational models of the human auditory system has also been shown to improve speech recognition accuracy in some difficult acoustical environments.

The research reported by Alejandro Acero in this monograph employs a third approach, that of transforming speech recorded from a single microphone in the application environment so that it more closely matches the important acoustical characteristics of the speech that was used to train the recognition system. His work builds on the older techniques of spectral subtraction and spectral normalization, which were originally developed to enhance the quality of degraded speech for human listeners. Spectral subtraction and spectral normalization were designed to ameliorate the effects of two complementary types of environmental degradation: additive noise and unknown linear filtering. The most important contribution of Acero's work is the development of a family of algorithms that *jointly* compensate for the effects of these two types of degradation. This unified approach to signal normalization provides significantly better recognition accuracy than the independent compensation strategies that had been developed in prior research. Joint compensation is facilitated by the use of cepstral-based speech representations that are able to incorporate the effects of additive noise and linear filtering, as well as the effects of nonlinear frequency warping, in a single affine transformation. The various algorithms described differ in their complexity, in their effectiveness in compensating for signal degradation, and in the type and amount of data needed to estimate compensation parameters.

The algorithms described in this monograph, such as codeword-dependent cepstral normalization (CDCN) and blind signal-to-noise-ratio cepstral normalization (BSDCN), have been shown to provide major improvements in recognition accuracy for speech systems in offices using desktop microphones, in automobiles, and over telephone lines. Although originally developed for speech recognition systems using discrete hidden Markov models, these algorithms are effective when applied to systems that use semi-continuous hidden Markov models as well. Real-time implementations have been developed for the compensation algorithms using workstations with onboard digital signal processors.

This monograph will be a valuable resource for several types of readers. It provides a comprehensive review and comparison of the major single-channel compensation strategies currently in the literature. It develops a unified cepstral representation that facilitiates joint compensation for the effects of noise, filtering, and frequency warping. Finally, it describes and explains the compensation algorithms that Acero has developed to compensate for these types of environmental degradation, and it provides the details needed to implement the algorithms.

Richard M. Stern

Acknowledgments

This work evolved from my thesis research at Carnegie Mellon University. First, I would like to thank Richard Stern, my thesis advisor, for his invaluable support, encouragement and guidance. His thorough writing abilities and scientific approach have helped me make this monograph a great deal better. Over the years we have developed a relationship that has contributed to making my stay at CMU a wonderful experience.

I am very grateful to the members of the CMU speech group. Raj Reddy has followed closely my research and suggested many good ideas. Kai-Fu Lee has helped me understand some of the mysteries in the SPHINX system. I owe Fu-hua Liu a great deal for the real-time implementation of CDCN and BSDCN, as well as for the controlled evaluation on these algorithms. Bob Weide patiently collected the databases I used. Shuji Morii's studies were the motivation for this work. Joel Douglas ran some of the experiments described in this monograph. I am grateful to Tom Sullivan and Yoshiaki Ohshima for their help. I also thank Pedro Moreno and Nobutoshi Hanai for proofreading drafts of this book.

I would also like to thank the Defense Advanced Research Projects Agency for supporting this work.

Finally I would like to thank my family and friends for their support. My brother was always there to give me good advice. My parents were always with me when I needed them and they have played a major role in these thesis years. They made big problems look insignificant and they definitely shared some of the success of this thesis. I will always be grateful to them.

Alejandro Acero

1 Introduction

Unconstrained automatic speech recognition (ASR) is a very difficult problem. Early ASR systems obtained a reasonable performance by artificially constraining the problem. Those systems were speaker dependent, and dealt with isolated speech for small vocabularies, while current speech recognizers produce higher performance under less constrained environments. SPHINX (Lee *et al.* [47] [46]), the system recently developed at CMU, is the first speaker-independent large-vocabulary continuous-speech recognition system. The technology has now reached the point where ASR systems may become useful in everyday applications although there are other considerations that must be addressed for speech to become a viable and facile man-machine communication medium:

- **Spontaneous speech**. For a system to be useful in a day-to-day application it has to accept spontaneous speech. Most current ASR systems deal with read speech only because spontaneous speech is often ungrammatical and ill-structured.
- **Semantics and pragmatics in speech understanding**. For a specific task, the perplexity, average number of words active at any given time, can be reduced substantially if semantic and pragmatic information is used in addition to a grammar.
- **Acoustical and environmental robustness**: ASR systems exhibit unacceptable degradations in performance when the acoustical environments used for training and testing the system are not the same. It would be desirable to have a system that works independently of recording conditions (using different rooms, microphones, noise levels).

In this monograph we focus on the last issue, the robustness with respect to changes in the acoustical environment, and we describe several algorithms that make SPHINX more *environment independent*.

1.1. Acoustical Environmental Variability and its Consequences

In this section we discuss the different causes of acoustical and environmental variability. There are some attributes of the environment that remain relatively constant through the course of an utterance such as the recording equipment, the amount of room reverberation, and the acoustical characteristics of the particular speaker using the system. Other factors, like the noise and signal levels, will be assumed to vary slowly compared to the rate at which speech changes.

Current HMM[1] recognizers usually focus on short-term attributes of the speech signal (over segments of about 20 ms that are called frames) while long-term properties are often ignored. We will discuss several different external factors that change the environment and therefore affect the performance of a speech recognition system: input level, additive noise, spectral tilt, physiological differences, and interference by the speech of other speakers (the cocktail party effect).

1.1.1. Input Level

Input level changes from utterance to utterance and even within the same utterance. Speakers normally do not speak with the same volume, distance from, and orientation to the microphone. Gain normalization cannot be done on a frame-by-frame basis since the input level is a long-term feature of speech.

[1]HMM, Hidden Markov Models, is the dominant technology in automatic speech recognition. See Chapter 2 for an overview.

1.1.2. Additive Noise

The performance of speech recognition systems degrades drastically when training and testing are carried out with different noise levels. When the signal-to-noise ratio (SNR)[2] is less than +10 dB the speech is severely corrupted and not even training and testing on the same noisy environment can ameliorate the problem.

In most cases background noise can be modeled as an additive stationary perturbation that is uncorrelated with the signal. We are excluding from this analysis non-stationary perturbations frequent in an office environment such as door slams, key strokes and other conversations. The stationarity of the background noise is another long-term property of our corrupted speech.

1.1.3. Spectral Tilt

Spectral tilt is the distortion observed in the spectrum when the speech signal is passed through an LTI (linear time-invariant) filter. In the log-spectral domain the net effect of spectral tilt is an additive offset to the transfer function. There are various sources of spectral tilt:

- **Speaker-specific spectral characteristics**. Differences in vocal tract physiology will lead to different long-term average frequency responses. Accurate speaker verification and identification systems have been built that use long-term spectral averages as their features.
- **Speech Styles**. It has been discovered recently [8] that spectral tilt is one of the major effects of the use of different speech styles (*e.g.* normal, soft, fast, and shouted speech, and speech spoken in the presence of noise [Lombard-effect speech]).
- **Room acoustics and reverberation**. The effect of reverberation can be modeled as a linear filter that depends on the geometry and materials of the room and the speaker location.
- **Recording equipment**. If a different microphone is used, the overall transfer function will change.

---

[2]Signal-to-noise Ratio is the ratio between the power of the signal and the power of the noise, and it is usually given in dB.

Spectral tilt is a major cause of performance degradation. The spectral distortion measures[3] used in speech recognition are severely affected by spectral tilt.

1.1.4. Physiological Differences

Differences in the physiology of the vocal tract will produce variability in the speech signal. Although formant frequencies are probably among the most invariant features in vowel discrimination, there is great variability among speakers due to differences in vocal tract sizes and shapes.

It is well known that the formant frequencies for a voiced sound will depend on the phonetic context (*short-term variability*) as well as on the anatomical characteristics of the speaker (*long-term variability*). Male voices exhibit lower formants than those of females, who in turn exhibit lower formants than those of children. The nominal resonant frequencies of the vocal tract depend strongly on the size of the vocal tract as well as on some other anatomical parameters.

1.1.5. Interference from Other Speakers

The interference that takes place when different speakers talk simultaneously is sometimes referred to as the cocktail party effect. In most speech recognition systems the input signal is modeled as an excitation (specifically an impulse train or white noise) driving a time-varying all-pole filter. Since this model assumes that only a single voice is present, interference from other speakers cause a dramatic degradation in recognition accuracy. While automatic recognition systems fail, humans do remarkably well in attending to and interpreting the speech of the desired speaker.

---

[3]Spectral distortion measures typically compute $\int_{-\pi}^{\pi} \{f[S_a(\omega)] - f[S_b(\omega)]\}^2 d\omega$ where the function $f(x)$ is x or a compressing function of x (the logarithm is a popular choice).

1.2. Previous Research in Signal Processing for Robust Speech Recognition

Robust speech recognitionis a young and rapidly growing field. Most of the early work toward robustness has been derived from classical techniques developed in the context of speech enhancement (Lim [51] offers a good summary of those techniques). In this section we describe different kinds of techniques used to enhance speech in the presence of additive noise.

1.2.1. Techniques Based on an Autoregressive Model

Autoregressive analysis (AR), is a set of techniques that assume that the signal spectrum can be represented by the all-pole transfer function

$$A(z) = \frac{G}{1 + \sum_{i=1}^{p} \alpha_i z^{-i}} \tag{1.1}$$

When such a signal is corrupted by additive noise, zeroes will appear as well as poles in the transfer function representing the signal. The goal of these approaches is to find the autoregressive (AR) parameters of the speech signal and the noise. Lim [50] proposed a speech enhancement technique that iterated back and forth between an estimate of the clean waveform obtained by Wiener filtering of the noisy speech and the AR parameters obtained from that clean waveform. Proof of the convergence of these techniques was later provided by Feder *et al.* [22]. Although this processing produced an increase in SNR, the processed speech was less intelligible, and the resulting estimates exhibited a large variance.

Hansen and Clements [33] [34] extended Lim's work by applying *inter-frame* and *intra-frame* constraints to reduce the variance of the estimates of the restored speech. When applied to speech recognition, they obtained a somewhat higher recognition rate. The work of Mansour and Juang [55] is also based on an AR model of speech.

The main problem with these approaches is that small inaccuracies of the all-pole model are magnified by this algorithm at high noise levels. Also, the restored speech may exhibit artificial resonances that do not correspond to real speech frames.

1.2.2. Techniques Based on Manipulation of Distortion Measures

A number of researchers have attempted to combat additive noise and spectral tilt by developing special distortion measures that are more robust to this type of variability. Gray *et al.* [31] presents an early study on different distortion measures for speech processing which favors the cepstral distance[4]. A more recent study was presented by Nocerino *et al.* [62].

Instead of using the Euclidean distance[5], some authors have proposed Mahalanobis distance[6] distances for cepstral feature vectors. Tokhura [82] showed that since different cepstral coefficients are uncorrelated with each other, it suffices to use a weighted Euclidean distance. The weights were the inverse of the variance of each cepstral coefficient obtained from the pooled data. For recordings on dial-up telephone lines this method proved to be quite effective.

Juang *et al.* [42] proposed the use of bandpass *liftering* [7], in which the cepstral coefficients were weighted by a raised sine function. They argued that reducing the contribution of low-order cepstral coefficients was beneficial because of their large variance, and that reducing the contribution of the highest-order cepstral coefficients improved the discrimination capabilities of ASR systems. Again this algorithm proved effective for dial-up telephone lines for digit recognition.

Other types of lifters were proposed by Hermansky [35], Itakura and Umezaki [40]. Junqua and Wakita [43] did a comparative study of lifters for robust recognition.

---

[4]The *cepstrum*, originally proposed by Bogert *et al.* [5], is the inverse Fourier transform of the logarithm of the spectrum. The cepstrum is a popular choice for speech recognition front-ends.

[5]The *Euclidean distance* between two vectors $\mathbf{x}$ and $\mathbf{y}$ is $d = [\sum (\mathbf{x}[k] - \mathbf{y}[k])^2]^{1/2}$.

[6]The *Mahalanobis distance* between vectors $\mathbf{x}$ and $\mathbf{y}$ is $d = (\mathbf{x} - \mathbf{y})^T \mathbf{C} (\mathbf{x} - \mathbf{y})$ where $\mathbf{C}$ is the inverse of the covariance matrix of $\mathbf{x}$ and $\mathbf{y}$.

[7]The word *liftering* comes from filtering with the first syllable reversed. The term was coined by Bogert *et al.* [5] in his work defining the cepstrum (spectrum backwards) as the inverse Fourier transform of the logarithm of the Fourier transform.

Mansour and Juang [55] proposed a set of distortion measures that gives consideration to the angle between two vectors, rather than just the norm of the difference. This distortion measure was very effective for combating dial-up telephone speech corrupted with additive synthetic noise for an alphanumeric isolated-word task.

Soong and Sondhi [77] proposed a frequency-weighted measure that accounted for the broadening of the peaks at low SNR. Goncharoff and Chandran [30] and Noda [63] did similar studies based on spectral warping.

Although relatively successful, one paradox of all these weighted Euclidean distortions is that they intend to combat distortions due to a shift in the mean (spectral tilt) by a variance normalization. Although these authors reported improvement on speech recorded over many different telephone lines over the case of no processing, we found in pilot experiments that Juang's liftering, Tokhura's weighted distance and Hermansky RPS distortion did not produce any useful improvement in recognition accuracy for our system using our database, perhaps because the frequency warping transformation used in SPHINX alters the variance of the cepstral coefficients.

1.2.3. The Use of Auditory Models

Since the human auditory system is very robust to changes in acoustical environments, some researchers have attempted to develop signal-processing schemes that mimic the functional organization of the peripheral auditory system. For example, Seneff [74] and Zue *et al.* [88], Hunt and Lefebvre [38], Ghitza [26] [27] and Cohen [10] all used models based on the human auditory system as front-end processors for their recognition systems.

While these methods appear to be able to produce some improvement in robustness for speech in noise, the mechanisms that enable this to occur are not well understood. In fact, Ghitza [27] found that the use of DFT[8] coefficients resulted in higher accuracy and robustness than the highly complex filter-bank. He found the power of the model for robustness to be in the non-linear stages. In later work, Hunt and Lefebvre [39] analyzed the

[8]Discrete Fourier Transform.

performance of different weighting schemes that used linear discriminant analysis (LDA) to combine heterogeneous speech features. In their experiments, a front-end based on LDA techniques attained a higher degree of robustness than the auditory model they described in [38] for the same task while being substantially more computationally efficient.

Cohen [10] designed a front-end based on some of the functions of the peripheral auditory system. When tested on the IBM TANGORA system, his front-end performed substantially better than a filter bank for speech recorded with a Crown PZM6A microphone.

While the use of models of the peripheral auditory system also provide a relative improvement for the case of mismatches in training and testing over the baseline-processing case, the reasons for this are again not well understood. Also, these techniques are typically very computationally expensive. Although we believe that great insight can be obtained by analyzing the functioning of the human auditory system, we believe that other alternative approaches are likely to be more practical in the shorter term.

1.2.4. Techniques Based on Short-Time Spectral Amplitude Estimation

A number of algorithms have been developed to estimate a clean speech spectrum from the noisy speech spectrum. Spectral subtraction techniques were introduced by Boll [6] and Berouti *et al.* [4] in the context of speech enhancement. The noise spectrum was subtracted from the corrupted speech spectrum. While increasing the SNR, the speech intelligibility was not improved.

Recent implementations of spectral subtraction for speech recognition systems include the work of Van Compernolle [12] [11] and Stern and Acero [78]. Van Compernolle improved the robustness of the IBM speech recognition system when a desk-top microphone was used. The feature vector in the IBM system consists of the outputs of an auditory model. He proposed the use of channel equalization with a special implementation of spectral subtraction in the logarithm domain. The work of Stern and Acero was similar, although it was based on cepstral parameters derived from an

LPC analysis.

Although *ad-hoc* techniques like spectral subtraction provide a moderate improvement, they are not derived with any optimum criterion. Porter and Boll [70] proposed to estimate DFT coefficients via MMSE[9] techniques, and this approach produced an improvement in performance over their original spectral-subtraction methods. They also showed that estimating the log-spectrum was more effective than estimating the spectrum directly as far as improving accuracy in the recognizer. The use of filter-bank outputs instead of DFT coefficients (Boll *et al.* [7]) resulted in even higher accuracy because of the smoothing effect across frequency.

The use of MMSE criteria provides more reliable estimates of the original speech signal at low SNR than the use of maximum likelihood estimation. These techniques are very effective, although they still suffer from the same plague that affects all approaches: at low SNR some of the processed speech does not represent legitimate speech.

1.2.5. Techniques Based on Mixture Densities

Some authors have very recently proposed to characterize the feature vectors input to speech recognizers by mixtures of Gaussian probability densities as a way of increasing noise robustness. Techniques based on short-time spectral amplitude estimation produced some frames at low SNRs that did not represent legitimate speech frames, because different frequency components were implicitly assumed to be independent of each other. From the plethora of density functions for which different components are not independent of each other, mixtures of Gaussian densities are a popular choice in the speech recognition community.

Ephraim *et al.* introduced MAP[10] [17] and MMSE estimates [18] for

---

[9]The minimum mean squared error criterion (MMSE) is used to derive estimates that minimize the expected squared error between the estimate and the random variable to be estimated.

[10]The maximum *a posteriori* criterion (MAP) selects the parameter that has the highest probability given the observed input. In this case, we select the most likely mixture component given the input frame.

speech enhancement with the aid of HMMs. Both speech and noise were modeled as AR processes whose parameters were computed via an EM[11] algorithm that maximized the likelihood across the whole utterance. The enhanced signal exhibited a substantial increase in SNR, although the process was very computationally expensive because the time waveform was reconstructed for every iteration.

Nadas *et al.* [59] used a mixture model of the outputs of an ear model in the TANGORA IBM speech recognition system. In their algorithm, the MIXMAX labeler, an MAP approach was used to select the most likely mixture component given the noisy speech frame. An EM algorithm was used to maximize the sum of Gaussian densities until convergence was reached. The MIXMAX labeler was shown to increase the robustness of TANGORA significantly but, since they assumed that the corrupted speech at each frequency band was the maximum of the noise and the clean speech at that band, low energy speech frames were not modeled accurately.

Erell and Weintraub [20] also proposed an approach that was based on a mixture model for filter-bank outputs. They reported improved performance over techniques that did not incorporate correlation between frequency bands. Since the parameter vector in DECIPHER, the SRI system, was the cepstrum, they computed an MMSE estimate that weighted the contributions from all the mixtures, instead of labeling directly as Nadas did. Their model for the noise degradation was more elaborate than that used by Nadas, but their system did not have the system integration that Nadas' had because they had to cluster the acoustic space in both the spectral domain and cepstral domain.

Nadas *et al.* [60] proposed an adaptive labeling scheme that applied a time-varying transformation to the input speech frames in an effort to minimize the accumulated VQ[12] distortion. Since this approach is directed towards normalizing different transfer functions as well as noise levels, the

---

[11]The EM, estimate-maximize, algorithm is an iterative algorithm to solve problems of maximum likelihood with incomplete data. The reader is referred to Appendix VI for details.

[12]Vector Quantization (VQ) is a procedure that assigns each input frame to the closest prototype vector from a finite set. The measure of proximity in SPHINX is the Euclidean distance. See Chapter 2 for details.

adaptive labeling increased the robustness of TANGORA with different microphones and recording conditions. However, analysis of convergence is hard for a constantly drifting process.

Varga *et al.* [83] proposed a model decomposition for speech and noise in which instead of estimating the speech in noise, the HMMs were modified to allow for the presence of noise via noise masking. In [84] Varga and Moore actually used an HMM for the noise and another HMM for the speech. The Viterbi algorithm involved a search in a 3D space in this case, and non-stationary noises could be modeled. Although this idea is very powerful, their model for noise degradation was not very accurate, as it was based on many hard thresholds. Also, the computational complexity was very high.

Furui [25] investigated an unsupervised speaker adaptation method that was based on hierarchical spectral clustering of cepstral coefficients. The goal of the approach was to minimize the VQ distortion between the target acoustic space and a universal acoustic space. He suggested that this technique could also be used to increase the robustness of the system in noisy environments.

1.2.6. Other Techniques

There are a number of different approaches to the problem of speech recognition systems that are robust to noise. Some of these other techniques include neural networks and the use of microphone arrays.

Tamura and Waibel [80] suggested the use of a neural network for speech enhancement that was trained to minimize the difference between noisy and clean waveforms. In [81] Tamura and Nakamura used different affine transformations between hidden layers that corresponded to different phonemes. Neural network technology may be promising for robust speech recognition too, but for that they should operate on the transformed domain, spectrum or cepstrum, and not on the waveform.

Some efforts are being pursued in enhancing speech by using a microphone array (Flanagan *et al.* [23], Silverman [76], Van Compernolle [13]). The goal of this approach is to develop a directivity pattern so that noise sources arriving from a different angle than the desired speech are attenuated. While microphone arrays need to be explored in the future, we believe that there is still room for improvement with a monophonic signal. In addition, these two approaches can complement each other.

While most previous efforts are geared towards suppressing stationary noise, some authors have focused on the difficult problem of speaker separation (Min *et al.* [56], Zelinski [87], Naylor and Boll [61]). These approaches work acceptably when the speech from both speakers is voiced and their pitch frequencies are far apart.

1.2.7. Discussion

One important factor in the selection of techniques to use was computational complexity, because we need to evaluate usefulness of speech input for personal computers. In this monograph we opted not to work with auditory models nor iterative techniques based on AR models since they both present a high degree of computational complexity. Furthermore, although these techniques provide an increased degree of robustness over the case of no processing, they have not been shown to be superior to the other techniques described.

At the other end of the continuum in computational complexity are the weighted distortion measures. Unfortunately, we have found in pilot experiments that the distortion measures proposed in the literature provided essentially no improvement over the case of standard Euclidean distance for our database evaluated with SPHINX.

Since in this study we were concerned with the capabilities and limitations of techniques that used only one microphone, we did not investigate microphone-array techniques. We also did not address the issue of non-stationary noise either because we consider it an extremely difficult problem.

In this monograph we will primarily explore approaches based on short-time spectral amplitude estimation and techniques that use mixture densities. They offer an attractive compromise between efficiency and accuracy.

1.3. Towards Environment-Independent Recognition

The goal of our research is to increase the robustness of the speech recognition systems with respect to changes in the environment. Since mismatches between training and testing conditions lead to a considerable degradation in performance, systems presently must be retrained for every different environment used. Even in the case of retraining, environments with a higher noise level suffer a loss of accuracy as compared to clean environments because some information is lost when noise is added. Going beyond that will require more sophisticated techniques.

A system can be labeled as *environment-independent* if the recognition accuracy for a new environment is the same or higher than that obtained when the system is retrained for that environment. Attaining such performance will be the goal of the work in this monograph.

In this section we will introduce the need for joint compensation of noise and equalization and an integrated approach for normalization in the cepstral domain.

1.3.1. Joint Compensation for Noise and Equalization

Most of the techniques described in the previous section to reduce the effects of mismatches between training and testing conditions deal only with additive noise. Nevertheless, we have also noted that *spectral tilt* can affect the performance of speech-recognition systems as well.

In this monograph we develop a set of algorithms to accomplish the *joint* normalization of noise and spectral tilt. Although some efforts have previously been made to combinethe two types of processing (*e.g.* Van Compernolle [11], Erell and Weintraub [20]), both phenomena were treated independently. We will show that there is indeed an interaction between these phenomena and that further benefit is obtained if joint normalization is performed.

We will present several algorithms that adapt to new acoustical environments by estimating the noise and spectral tilt from input data. For this we use the concept of a universal acoustic space that is transformed to match the acoustic space of the current environment. We show that a few seconds of speech are sufficient to adapt to a different acoustical environment.

1.3.2. Processing in the Cepstral Domain: A Unified View

While successful noise-suppression algorithms operate in the frequency domain, most successful continuous speech recognizers operate in the cepstral domain. In this monograph we will describe algorithms that perform the noise suppression in the cepstral domain, so that a larger degree of integration can be achieved for cepstral-based systems. In fact, the level of integration is so high that the algorithms can be implemented in an extremely efficient manner.

We show that compensation for noise and spectral tilt can be achieved by an additive correction in the cepstral domain. In addition to this bias compensation, a variance normalization is also possible. Finally, we propose a frequency normalization operation that can be expressed as a matrix multiplication operation.

We show in this monograph that conditioning on the instantaneous SNR[13] is advantageous. We show that the instantaneous SNR captures a great deal of the information needed to perform the compensation.

Although frequency normalization may appear to be outside the scope of the problems of robustness to changes in the environment, we show that using the processing in the cepstral domain together with the concept of normalization of the acoustic space allow us to perform some adaptation to the long-term characteristics of the speaker. In this monograph we present a novel method to accomplish this via adaptive warping of the frequency axis.

---

[13]We define the instantaneous SNR as the ratio of the short-time energy of the speech at a certain time over the average energy of the noise. For this work, we assume the noise to be stationary, so that the noise energy is supposed to be constant and can be estimated from previous frames.

1.3.3. Measuring Performance Evaluation

Although many of the techniques described in the previous section are successful at some level, it has been difficult to compare them with each other because there does not exist a standard corpus for algorithm comparisons in the field of noise robustness like the one developed by DARPA for research in continuous speech recognition. The way many authors evaluate their algorithms is by comparing recognition accuracies for different noise levels for a system that has been trained with clean speech, with and without processing for robustness. While these measurements may demonstrate the effectiveness of a given method, they do not provide a basis for comparison with other authors' work because different authors use different tasks and noise levels. With a few exceptions, most authors do not consider the accuracy of the system when it is trained and tested on the noisy speech. We believe that this is an important benchmark for the evaluation of algorithms. To this end, we recorded a training database stereophonically using two different microphones: a close-talking microphone and a desk-top microphone.

Another common characterization of noisy speech databases is that of SNR. Although under some circumstances SNR can provide a good estimate of the degree of difficulty of a speech database that has been recorded in the presence of white noise, it does not characterize the database when the noise is colored or when the speech has been passed through a filter that has altered the frequency response, as is the case for real environments. We will use the simple average of speech and noise spectra as a characterization of a stereo database.

1.4. Monograph Outline

Chapter 2 describes the structure of the speech recognitionsystem and the databases. Chapter 3 reviews some of the specifics of implementations of spectral subtraction and equalization, as well as other standard techniques described in the literature.

Chapter 4 presents the need for a joint compensation of noise and spectral tilt. Two algorithms will be presented: one in the frequency domain and the other in the cepstral domain. The latter introduces the idea of a normalization that depends exclusively on the instantaneous SNR of the input speech and is the basis for the *SNR-Dependent Cepstral Normalization* (SDCN) algorithm.

Chapter 5 introduces the *Codeword-Dependent Cepstral Normalization* (CDCN) algorithm to solve the problem of changes in the environment by using the idea of normalization of the acoustic space. CDCN also exhibits a better behavior for frames with low SNR.

Chapter 6 introduces several algorithms that build on SDCN and CDCN in an effort to obtain more accurate and efficient algorithms. We present the *Interpolated SDCN* (ISDCN) algorithm, the *Blind SDCN* (BSDCN) algorithm and the *Fixed CDCN* algorithm as an evolution of the algorithms in Chapters 5 and 4.

In Chapter 7 we describe a method to perform frequency normalization within the context of cepstral processing in SPHINX. We use the bilinear transform to find the warping parameter of the frequency axis that minimizes the VQ distortion. Finally, Chapter 8 contains a summary of results, and Chapter 9 contains our conclusions and suggestions for future work.

2 Experimental Procedure

In this chapter we will give an overview of the SPHINX system, describe the database we used in this work, SNR characterizations, and the baseline results. It is important to note that in this study we were not concerned with elevating the absolute performance for a specific task but rather with comparing the relative merits of different algorithms.

2.1. An Overview of SPHINX

The SPHINX system was developed at CMU by Lee *et al.* [48] [46], the pioneer in speaker-independent large-vocabulary continuous-speech recognition. We will briefly describe the different blocks that compose the system, with greater emphasis in the signal processing and vector quantization, since these are aspects that this monograph concentrates on.

A block diagram of the front-end stages of SPHINX is shown in Figure 2-1.

2.1.1. Signal Processing

All speech recognition systems use a parametric representation rather than the waveform itself. Typically the parameters carry information on the envelope of the spectrum. SPHINX uses frequency-warped LPC cepstrum as its parameter set, that are computed as follows:

- Speech is digitized at a sampling rate of 16 kHz.
- A Hamming window of 320 samples (20 ms) is used every 10

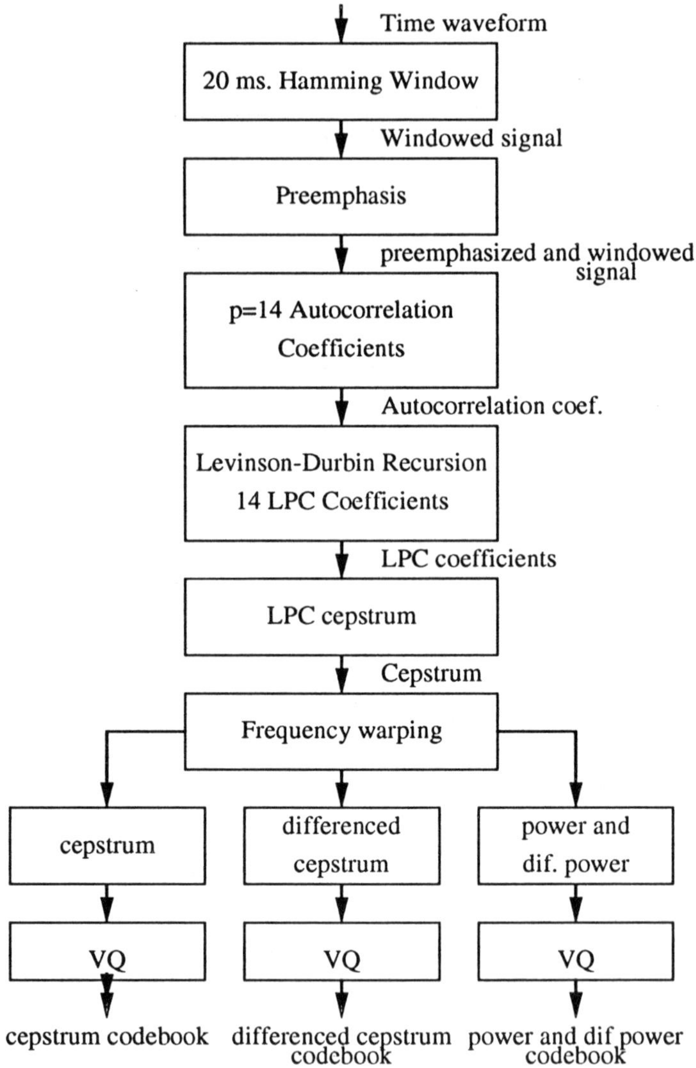

Figure 2-1: Block diagram of SPHINX front-end.

ms.

- A preemphasis filter $H(z) = 1 - 0.97\,z^{-1}$ is applied.

- 14 Autocorrelation coefficients are computed.
- 14 LPC coefficients are derived from the Levinson-Durbin recursion.
- 32 LPC cepstral coefficients are computed using the standard recursion
- These cepstral coefficients are frequency warped by using the bilinear transform producing 12 warped LPC cepstral coefficients.
- To account for differences in overall input level, the maximum value of the power g[14] in the utterance was subtracted for all frames so that after power normalization the maximum g is 0.

Although adjacent frames are indeed correlated with each other, the SPHINX system assumes that every frame is statistically independent of the rest. In addition to the static information provided by the cepstrum, SPHINX also uses dynamic information represented by the first order difference of cepstral vectors:$\mathbf{d}_i = \mathbf{x}_{i+2} - \mathbf{x}_{i-2}$. See Appendix II for a more detailed description of this processing.

2.1.2. Vector Quantization

Vector quantization (VQ) (Gray [32])is a data reduction technique that maps a real vector onto a discrete symbol. Although it was originally proposed for speech coding, it has gained a great deal of popularity in speech recognition recently. A vector quantizer is defined by a codebook and a distortion measure:

- The discrete alphabet, called *codebook*, contains L vectors, and it is a quantized representation of the vector space.
- The distortion measure estimates the degree of proximity of two vectors. An input vector is mapped to a symbol of this alphabet by choosing the *closest* codebook vector. In SPHINX the distortion measure used is the *Euclidean* distance.

The version of SPHINX that we used for this work has three different codebooks: one for the cepstrum, one for the first difference of cepstral

[14] g is the zeroth order cepstral coefficient.

vectors and the last one for power and the first difference of the power. By having each codebook contain 256 vectors, every frame of speech is condensed to 3 bytes. The distortion measure used is the Euclidean distance. The *prototype* vectors are estimated via a hierarchical clustering algorithm similar to the K-means algorithm developed by Linde *et al.* [52], which is an approximate maximum-likelihood method.

2.1.3. Hidden Markov Models

Hidden Markov Models (HMM), the dominant technology in continuous speech recognition, constitutes the recognition engine used in SPHINX. Rabiner and Juang [72] present a good review of HMMs, Picone [69] offers a summary of HMM applications to speech recognition.

Briefly, an HMM is a collection of states connected by transitions. Each transition carries two sets of probabilities:

- A transition probability which provides the probability for taking a transition from one state to the next, and
- An output probability density function (*pdf*), which defines the conditional probability of emitting each output symbol from a finite alphabet given that that transition is taken.

HMMs have become a widely-used approach for speech recognition due to the existence of maximum likelihood techniques to estimate the parameters of the models and algorithms that efficiently find the most likely state sequence.

2.1.4. Speech Units

Although the natural choice for a speech unit is the word, it is an impractical one due to the large number of words and the fact that numerous repetitions of each are needed to train the system adequately. A common speech unit is the phoneme, typically a 3-state left-to-right HMM, because it is easily trainable and because there are a small number of them (45 in an early version of SPHINX). Since the same phoneme in different contexts can be very different, SPHINX uses generalized *triphone* models as a way to model both the left and the right context.

Models for words are obtained by concatenating the models of the appropriate generalized triphones. It was found that using *function-word-dependent phones* helped improve recognition accuracy of function words such as *the, a, in, with* that occur frequently and are poorly articulated.

Finally the model for a complete utterance is obtained by a concatenation of the word models. A grammar can be incorporated in the HMM as another network whose nodes are words with different transition probabilities.

2.2. The Census Database

Although the bulk of the early research using the SPHINX system at Carnegie Mellon has made use of the well-known Resource Management database, we elected to use a different database, the census database, for our evaluations of signal processing. There are two reasons for this:

- The Resource Management database, with its large vocabulary size and many utterances, required about a week to train satisfactorily, which was excessively long since the entire system had to be retrained each time a new signal-processing algorithm was introduced.
- We specifically wanted to compare simultaneous recordings from close-talking and desk-top microphones in our evaluations. We believe that it is important to evaluate speech-recognition systems in the context of natural acoustical environments with natural noise sources, rather than using speech that is recorded in a quiet environment into which additive noise and spectral tilt are artificially injected.

We will now specify the speaker population, the database contents, the environment in which it was recorded and the recognition system we used to evaluate our algorithms.

2.2.1. Speaker Population

Since the system is to be speaker-independent, a large number of speakers are needed to train SPHINX. The census database has two disjoint segments:

- The training segment of the census database contains utterances from 74 speakers (53 male and 21 female).
- The testing segment contains utterances from 10 different speakers (7 male and 3 female) not present in the training.

All the speakers were selected from staff and students at CMU. The male/female ratio in the database reflects that of the general CMU population.

2.2.2. Database Contents

The database consisted of strings of letters, numbers, and a few control words, that were naturally elicited in the context of a task in which speakers spelled their names, addresses, and other personal information, and entered some random letter and digit strings. Specifically, each speaker read :

- 5 alphanumeric ("an") utterances that contained letters, digits and some control words (enter, erase, go, help, no, repeat, rubout, start, stop, yes) randomly. Some sample utterances are **N-S-V-H-6-T-49** and **ENTER-4-5-8-2-1**.
- 9 census ("cen") utterances containing respectively last name, first name, street number, street name, city, zip code, home phone number, birth date, and Social Security Number. Some sample utterances are **R-O-B-E-R-T** and **P-I-T-T-S-B-U-R-G-H**.

Since some of these utterances were discarded due to bad recordings, the total number of utterances for the training database is 1018. The testing segment of the database contains 140 utterances.

2.2.3. Alphanumeric Database

A total of 104 vocabulary items appeared in the vocabulary, of which 41 were uttered fewer than 10 times. No grammar was used in any of the experiments. This census task presents a greater degree of difficulty than the Resource Management task because:

- The perplexity[15] is larger. The census database with no grammar has a perplexity of 104, while the perplexity of the word-pair grammar used in the resource management database is 60.
- The census vocabulary exhibits greater intrinsic acoustic confusability than the resource management vocabulary, because short words used in the census task such as the ones in the E-set (**B**, **C**, **D**, **E**, **G**, **P**, **T**) present a higher degree of acoustical similarity than the longer words in resource management.

2.2.4. The Environment

The census database was recorded simultaneously *in stereo* using both the Sennheiser HMD224 close-talking microphone that has been a standard in previous DARPA evaluations, and a desk-top Crown PZM6fs microphone. The recordings were made in one of the CMU speech laboratories (the "Agora" lab), which has high ceilings, concrete-block walls, and a carpeted floor. Although the recordings were made behind an acoustic partition, no attempt was made to silence other users of the room during recording sessions, and there is consequently a significant amount of audible interference from other talkers, key clicks from other workstations, slamming doors, and other sources of interference, as well as the reverberation from the room itself. Since the database was limited in size, it was necessary to perform repeated evaluations on the same test utterances.

2.2.5. The Recognition System

We also performed these evaluations using a more compact and easily-trained version of Sphinx with only 329 triphone models, omitting such features as duration, function-word and function-phrase models, between-word triphone models, and corrective training. We were willing to tolerate the somewhat lower absolute recognition accuracy that this version of Sphinx provided because of the reduced time required by the training process. Using the census database, the more compact Sphinx system, and DEC 3100

[15]*Perplexity* is an information theoretic measure of the amount of constraint imposed by a finite-state grammar. If no grammar is used, the perplexity coincides with the size of the vocabulary. If a grammar is used to restrain the search space, the perplexity will be lower than the size of the vocabulary. In general higher perplexity tasks produce higher error rates.

workstation, we were able to reduce the training time to the point that an entire train-and-test cycle could be performed in about 10 hours.

2.3. Objective Measurements

In this section we will summarize some definitions of Signal-to-Noise Ratio (SNR) given in the literature (See Jayant and Noll [41]) and present a new measurement that we believe provides more information about the environment.

Since the speech present in both microphone recordings was the same (as the database was recorded in stereo simultaneously), the main differences between them would be the noise levels and general tilt of the spectrum. We will see that conventional SNR figures of merit can only measure the difference in overall noise power, while a more detailed representation may be desired.

2.3.1. Measurements for Stationary Signals: SNR

SNR is defined as the ratio between the signal variance and the noise variance for a given signal $x[n]$. The noise variance is typically estimated as an average over a segment of a waveform that contains only noise samples.

If the speech database contains silent segments and we don't exclude them in the calculation of the signal power, the SNR figure will depend on how much silence our database contains, which is clearly undesirable for any objective measure. The fact that noise samples should be excluded in the computation brings up the issue of speech/silence discrimination.

The problem of discriminating speech and noise can be considered as a pattern recognition problem with two classes. Independently of what algorithm we use for this, the probability of labeling speech as noise and vice versa will be non-zero. The problem will become more severe as the noise level gets closer to the signal level, since the difference in means between the two distributions will be smaller.

In the case of our stereo database, the Sennheiser HMD224 close-talking microphone (CLSTK) exhibited an SNR of 38.4 dB whereas the

Crown PZM6fs (CRPZM) had an SNR of 19.7 dB. A 20 ms Hamming window was used in the calculations.

2.3.2. Measurements for Nonstationary Signals: SEGSNR and MAXSNR

The above defined SNR is an adequate characterization of a channel if both signal and noise are stationary. For the time being we will be concerned with stationary noise, but it is well known that the speech signal is clearly *non-stationary*. By using the SNR as a measure, the high energy segments of the signal are dominating the computation. If this average is computed for a complete database, the overall SNR will be dominated by the speakers speaking most loudly.

To alleviate the fact that the speech signal is nonstationary and that the same amount of noise has different perceptual values depending on the ambient signal level, Noll [41] proposed the *segmental* SNR. The segmental SNR is based on a log-weighting that converts component SNR values to dB prior to averaging, so that very high SNR segments do not camouflage other segments with low SNR. A window has to be used in which speech is considered stationary (in our case a 20 ms Hamming window) as a basis for short-time integration. For our database the CLSTK and CRPZM exhibited a SEGSNR of 31.2 dB and 16.0 dB respectively.

Another measure proposed in the literature is SNRMAX, the maximum SNR in dB. This peak SNR has the advantage that it is not affected by the amount of silence the database has. On the other hand, just like when measuring maxima of *pdf*s, it is necessary to specify how the length of the window used (from one utterance to the whole database) and how the maximum is computed. We computed the SNRMAX as the average over all utterances in the test set of the maximum SNR. The results for the CLSTK and CRPZM were 50.3 dB and 29.6 dB respectively.

2.3.3. Frequency-Weighted SNR

This refinement recognizes that noise in certain frequency bands is less harmful than that in other bands of an input signal, and that signal in certain bands contributes more to intelligibility/recognition rate than in other bands.

These measures have been extensively used in telephony:

- The *C-message weighting* function for speech represents the frequency response of the 500-type telephone set, as well as the hearing characteristics of the average telephone user.
- The *psophometric weighting* function, the European CCITT standard, is very similar to the C-message weighting function.
- For the *Articulation Index* used in the early speech work, the speech signal is observed in 20 sub-bands approximately distributed according to a Bark scale[16],with each band contributing equally to intelligibility. Component SNR values in dB (limited to a maximum allowable of 30 dB) are averaged:

$$AI = 0.05 \sum_{i=1}^{20} [\ \min\{SNR_i, 30\} / 30] \quad (2.1)$$

If, for example, the recordings include a 60 Hz hum, the SNR figure will be low although the recognition rate can still be high if those frequencies are filtered out. A measurement that would exclude those frequency bands would give a better estimate of the difficulty of the database from the standpoint of recognition accuracy.

The SPHINX system uses a technique similar to the *AI* based on frequency warping and a preemphasis filter that improves its accuracy. We argue that it is more meaningful to use a measure that uses this sort of frequency-weighted SNR in its calculation.

2.3.4. A Proposed Solution: Average Speech and Noise Spectra

Since no single SNR measure will tell us anything about a tilt in the spectrum or about how the speech and noise energies are distributed across frequencies, we introduced a way of characterizing the databases that presents the results as a function of frequency.

To discriminate speech/non-speech frames we used a simple threshold on the zeroth cepstral component *g*, which is closely related to the energy of

---

[16]The *Bark scale* is a warping of the frequency axis that is intended to better represent the frequency selectivity of the peripheral auditory system than a linear frequency scale.

the frame. The minimum g_{min} in the utterance was computed and all the frames below $g_{min} + 0.8$ were considered noise while all the frames above $g_{min} + 1.5$ were considered speech. The frames between these two thresholds are not included in the calculation. The cepstrum vectors of all frames classified as noise are averaged together, and so are the frames classified as speech. The DFTs of the resulting cepstral averages for the census database are plotted in Figure 2-2 for the Sennheiser HMD224 (CLSTK) and Crown PZM6fs (CRPZM), and the curves were normalized so that the noise level was 10 dB. The thresholds for classifying noise and speech were chosen empirically. The curves in Figure 2-2 computed by this method were not very sensitive to variations of the thresholds around the selected values. The classification done by this method was conservative in the sense that the system seldom misclassified speech as noise or vice versa. The price to pay was to have a region in which the frame was not classified as either noise or speech, but this was not important in this case as only an average is needed. In Chapter 6 we described a more elaborate method for noise estimation (Van Compernolle [12]) that performed similarly.

By comparing the curves in Figure 2-2, it can be seen that speech is about 25 dB above the noise level using the close-talking Sennheiser microphone. The signals from the Crown PZM, on the other hand, exhibit an SNR of less than 10 dB for frequencies below 1500 Hz and about 15 dB for frequencies above 2000 Hz. Furthermore, the response of the Crown PZM exhibits a greater spectral tilt than that of the Sennheiser, perhaps because the noise-canceling transducer on the Sennheiser also suppresses much of the low-frequency components of the speech signal. The so-called spectral tilt is the difference between the speech spectrum of the two microphone recordings.

The separation between curves in Figure 2-2 is smaller in magnitude than the SEGSNR, because these average spectra are computed by averaging the log-energies of all frequency bands rather than the energies themselves, as used in the SEGSNR calculation.

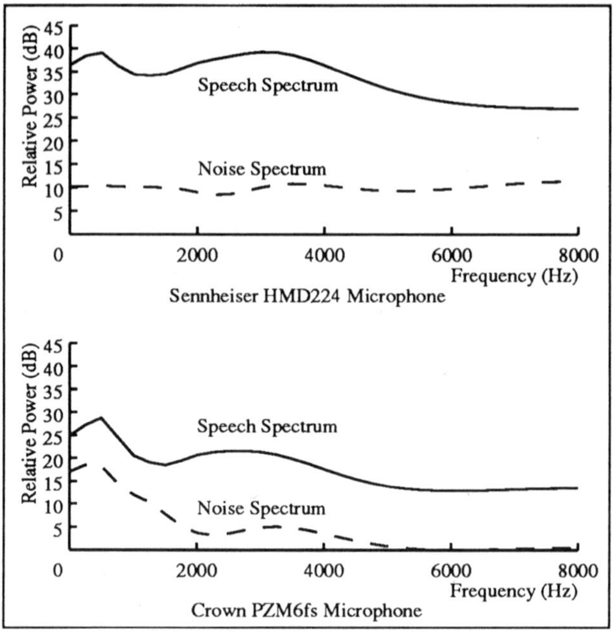

Figure 2-2: Average speech and noise spectra from the Alphanumeric database obtained using the headset-mounted Sennheiser HMD224 Microphone and the Crown PZM6fs microphone. The separation of the two curves in each panel provides an indication of signal-to-noise ratio for each microphone as a function of frequency. It can also be seen that the Crown PZM6sf produces greater spectral tilt.

2.3.5. Discussion of SNR Measures

In summary, the characterization of a speech database should not depend on aspects of the speech signal that are irrelevant such as

- Silent Periods and Background Noise. The objective measurement should reflect the *quality* of the recording, which does not depend on the length of the silent periods or background noise.
- Bandpass filtering. The level of an unfiltered speech signal mainly depends on the frequency components below 500 Hz while those frequencies only carry a part of the information.

High-pass filters with cut-off frequencies of around 300 Hz are used in telephone communications.

We would expect that the acoustical characterization of some speech material will give some indication of how well a speech recognition system or a human can perform on it. In other words, the characterization should be correlated with the recognition rate and intelligibility. The use of average spectra provides a better indication of the difficulty of the database than measurements such as the SEGSNR because :

- It provides the SEGSNR for every frequency (noise in some frequencies is more harmful than in others), as opposed to averaging the SNR across all frequencies.
- It provides an indication of the tilt in the spectrum, which can degrade performance significantly, even for the same SEGSNR.

	CLSTK	**CRPZM**
MAXSNR	50.3 dB	29.6 dB
SNR	38.4 dB	19.7 dB
SEGSNR	31.2 dB	16.0 dB
AVGSPT	24.4 dB	13.3 dB

Table 2-1: Analysis of different SNR measures for the census database. In all cases a 20 ms Hamming window was used. All the figures are computed as the average in dB across all the utterances in the database of different measures: maximum signal energy in the utterance for MAXSNR, the average signal energy for SNR, the average log-energy of the signal for SEGSNR. AVGSPT is the average separation of curves in Figure 2-2

Table 2-1 shows the different SNR measures described for the census database. The differences between SNR, SEGSNR and AVGSPT are whether average of energies or log-energies are used across time and/or frequency.

2.4. Baseline Recognition Accuracy

We first consider the "baseline" recognition accuracy of the SPHINX system obtained using the two microphones with the standard signal processing routines. Table 2-2 summarizes the recognition accuracy obtained by training and testing using each of the two microphones. Recognition accuracy is reported using the standard DARPA scoring procedure reported by Pallett [65], withpenalties for insertions and deletions as well as for substitutions. It can be seen that training and testing on the Crown PZM produces an error rate that is 60% worse than the error rate produced when the system is trained and tested on the Sennheiser microphone. When the system is trained using one microphone and tested using the other, however, the performance degrades to a very low level.

Hence we can identify two goals of signal processing for greater robustness: we need to drastically improve the performance of the system for the "cross conditions", and to elevate the absolute performance of the system when it is trained and tested using the Crown PZM.

	Test **CLSTK**	Test **CRPZM**
Train **CLSTK**	85.3 %	18.6%
Train **CRPZM**	36.9%	76.5%

Table 2-2: Baseline recognition rate of the Sphinx system when trained and tested on the census vocabulary using each of the two microphones.

2.4.1. Error Analysis

In order to better understand why performance degraded when the microphone was changed from the CLSTK to the CRPZM, even when the CRPZM was used for training as well as testing, we analyzed the errors that occurred in the CRPZM that did not occur in the CLSTK. We studied the spectrograms and listened carefully to all utterances for which training and testing with the CRPZM produced errors that did not appear when the system was trained and tested on the CLSTK. The estimated causes of the "new" errors using the CRPZM are summarized in Table 2-3. Not too surprisingly, the major consequence of using the CRPZM was that the effective SNR was lowered. As a result, there were many confusions of silence or noise segments with weak phonetic events. These confusions accounted for some 55 percent of the additional errors, with crosstalk (either by competing speakers or key clicks from other workstations) identified as the most significant other cause of new errors.

Type of error	Percent errors
Weak-event insertion	41.5
Weak-event deletion	13.2
Crosstalk	20.0
Others	25.3

Table 2-3: Analysis of causes of "new" errors introduced by use of the Crown PZM microphone.

2.5. Other Databases

Although the main database used in this work is the census database, we collected recordings using other microphones as well. In all cases the system was trained using the Sennheiser HMD224 (CLSTK). The "second" microphones (with which the system was *not* trained) were:

- The Crown PCC160 desk-top phase-coherent cardioid microphone (CRPCC160). (This is the new DARPA "standard" desk-top microphone.)
- An independent test set using the Crown PZM6fs.
- The Sennheiser 518 dynamic cardioid, hand-held microphone (SENN518).
- The Sennheiser ME80 electret supercardioid stand-mounted microphone (SENNME80).
- An HME lavalier microphone that also used an FM receiver (HME).

We recorded 140 utterances from 10 speakers for every microphone above and they were used in evaluating some algorithms after they had been developed.

We now summarize the salient acoustical attributes of these databases. The microphone recordings can be ranked according to decreasing SNR: the standard Sennheiser HMD224 (CLSTK), the Sennheiser 518, the Sennheiser ME80, HME FM, Crown PCC160 and Crown PZM6fs. Recordings with the Crown PCC160, Crown PZM6fs and HME FM exhibit considerable spectral tilt. A noticeable coloration of the noise spectrum is present for the Crown PZM6sf and the HME FM.

2.5.1. Sennheiser HMD224 - Crown PCC160

The average spectra for stereo recordings using the Sennheiser HMD224 and the Crown PCC160 are shown in Figure 2-3, and the SNR measurements are presented in Table 2-4. The baseline accuracy for the Sennheiser HMD224 and the Crown PCC160 were 82.4% and 70.2% respectively.

	CLSTK	CRPCC160
MAXSNR	52.8 dB	36.1 dB
SNR	41.2 dB	26.0 dB
SEGSNR	33.4 dB	20.9 dB

Table 2-4: Comparison of MAXSNR, SNR and SEGSNR measurements for the alphanumeric database recorded with the Sennheiser HMD224 and the Crown PCC160.

2.5.2. Sennheiser HMD224 - Crown PZM6fs

The average spectra for stereo recordings using the Sennheiser HMD224 and the Crown PZM6fs are shown in Figure 2-4, and the SNR measurements are presented in Table 2-5. The baseline accuracy for the Sennheiser HMD224 and the Crown PZM6fs were 84.8% and 41.8% respectively.

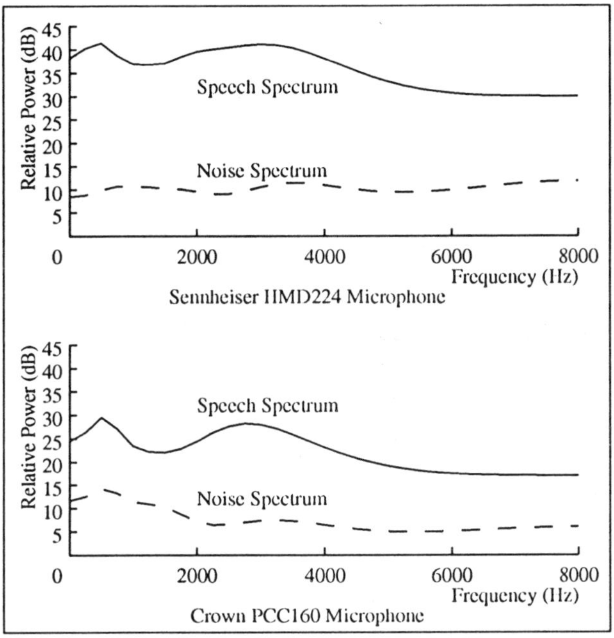

Figure 2-3: Average speech and noise spectra from the Alphanumeric database obtained using the headset-mounted Sennheiser HMD224 Microphone and the Crown PCC160 microphone. The separation of the two curves in each panel provides an indication of signal-to-noise ratio for each microphone as a function of frequency.

	CLSTK	**CRPZM6fs**
MAXSNR	53.0 dB	33.4 dB
SNR	41.5 dB	23.4 dB
SEGSNR	33.6 dB	18.9 dB

Table 2-5: Comparison of MAXSNR, SNR and SEGSNR measurements for the alphanumeric database recorded with the Sennheiser HMD224 and the Crown PZM6fs.

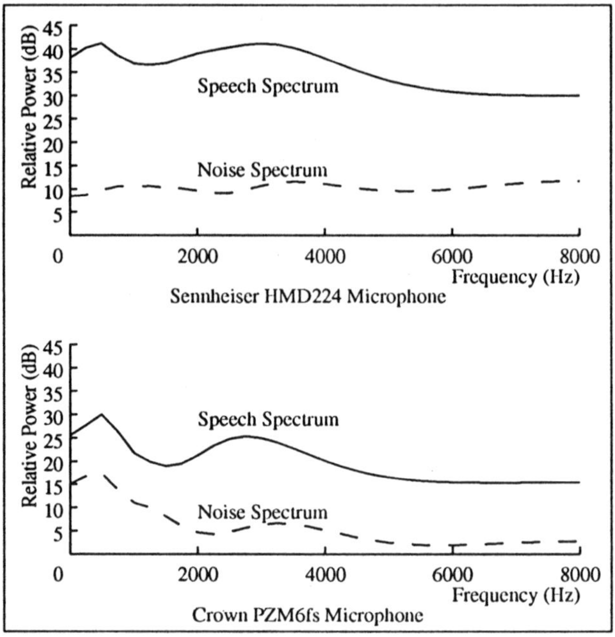

Figure 2-4: Average speech and noise spectra from the Alphanumeric database obtained using the headset-mounted Sennheiser HMD224 Microphone and the Crown PZM6fs microphone. The separation of the two curves in each panel provides an indication of signal-to-noise ratio for each microphone as a function of frequency.

2.5.3. Sennheiser HMD224 - Sennheiser 518

The average spectra for stereo recordings using the Sennheiser HMD224 and the Sennheiser 518 are shown in Figure 2-5, and the SNR measurements are presented in Table 2-6. The baseline accuracy for the Sennheiser HMD224 and the Sennheiser 518 were 87.2% and 84.5% respectively.

	CLSTK	SENN518
MAXSNR	51.8 dB	44.8 dB
SNR	40.4 dB	34.0 dB
SEGSNR	32.6 dB	27.4 dB

Table 2-6: Comparison of MAXSNR, SNR and SEGSNR measurements for the alphanumeric database recorded with the Sennheiser HMD224 and the Sennheiser 518.

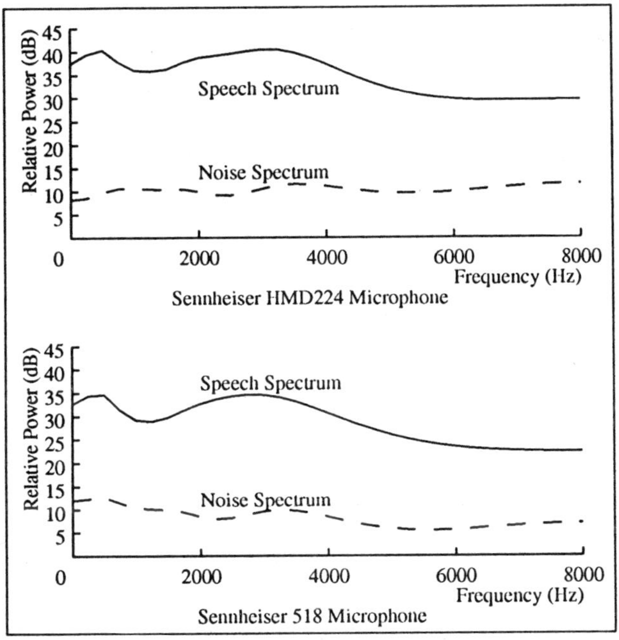

Figure 2-5: Average speech and noise spectra from the Alphanumeric database obtained using the headset-mounted Sennheiser HMD224 Microphone and the Sennheiser 518 microphone. The separation of the two curves in each panel provides an indication of signal-to-noise ratio for each microphone as a function of frequency.

2.5.4. Sennheiser HMD224 - Sennheiser ME80

The average spectra for stereo recordings using the Sennheiser HMD224 and the Sennheiser ME80 are shown in Figure 2-6, and the SNR measurements are presented in Table 2-7. The baseline accuracy for the Sennheiser HMD224 and the Sennheiser ME80 were 83.7% and 71.4% respectively.

	CLSTK	SENNME80
MAXSNR	52.2 dB	41.2 dB
SNR	40.7 dB	30.0 dB
SEGSNR	33.0 dB	23.2 dB

Table 2-7: Comparison of MAXSNR, SNR and SEGSNR measurements for the alphanumeric database recorded with the Sennheiser HMD224 and the Sennheiser ME80.

2.5.5. HME FM - Crown PCC160

The average spectra for stereo recordings using the HME FM and the Crown PCC160 are shown in Figure 2-7, and the SNR measurements are presented in Table 2-8. The baseline accuracy for the HME FM and the Crown PCC160 were 55.9% and 56.3% respectively.

2.6. Summary

In this chapter a description of the experimental procedures to be used in the monograph is presented. We are fortunate to be able to use the state-of-the-art recognition engine. We described in more detail the front-end in SPHINX, and especially the cepstrum and VQ stages, since they will be central to the work in the rest of the monograph.

After providing an overview of the SPHINX system, we proceeded to describe the census database for speaker-independent continuous speech that will be used in this monograph. The database was recorded in stereo simultaneously with two different microphones: a standard close-talking microphone and a desk-top one. This allow us to compare results in a more

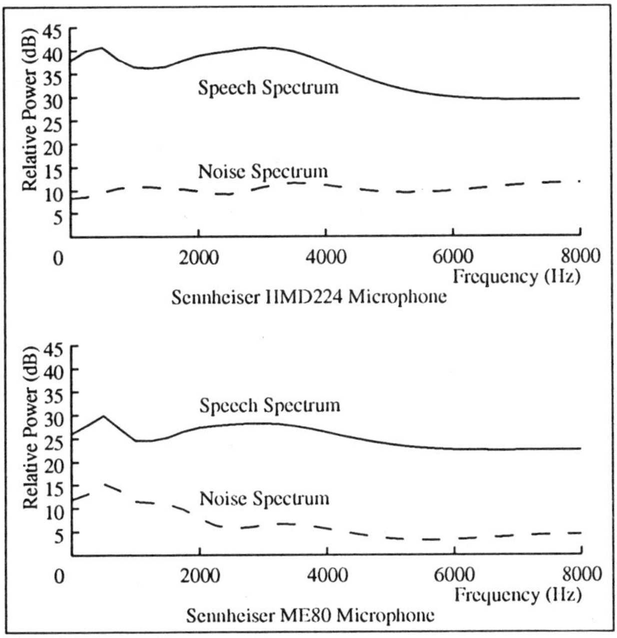

Figure 2-6: Average speech and noise spectra from the Alphanumeric database obtained using the headset-mounted Sennheiser HMD224 Microphone and the Sennheiser ME80 microphone. The separation of the two curves in each panel provides an indication of signal-to-noise ratio for each microphone as a function of frequency.

	HMEFM	**CRPCC160**
MAXSNR	36.7 dB	35.1 dB
SNR	26.4 dB	25.1 dB
SEGSNR	21.2 dB	20.1 dB

Table 2-8: Comparison of MAXSNR, SNR and SEGSNR measurements for the alphanumeric database recorded with the HME FM and the Crown PCC160.

direct way.

Several possible characterizations of the database were proposed as a

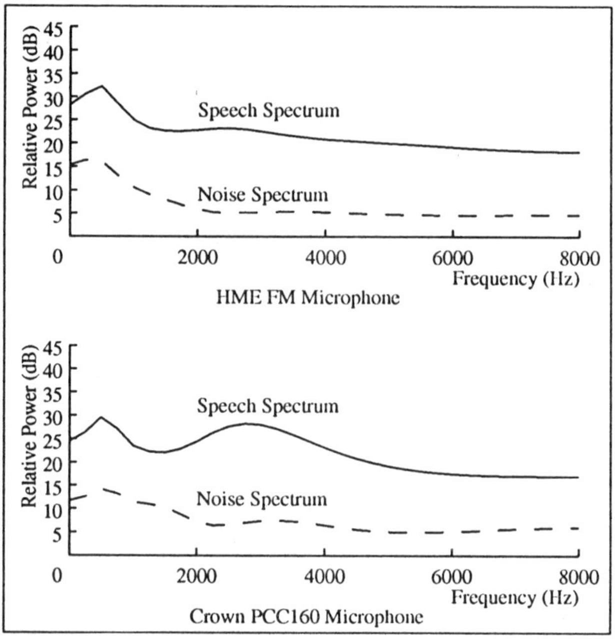

Figure 2-7: Average speech and noise spectra from the Alphanumeric database obtained using the HME FM Microphone and the Crown PCC160 microphone. The separation of the two curves in each panel provides an indication of signal-to-noise ratio for each microphone as a function of frequency.

measure of its difficulty. While standard signal-to-noise ratio calculations provide a useful measurement, the proposed average spectra is a more detailed and useful characterization.

We then presented the baseline performance of the census database with SPHINX. The main results were that mismatches in training and testing conditions lead to a considerable degradation in performance. The motivation here is to increase the robustness of speech recognition systems that are tested under a number of different acoustical environments.

Finally, additional databases with different sets of microphones were recorded in stereo. The average spectra, SNR calculations and baseline performance were computed for all of them.

3 Frequency Domain Processing

In this chapter we review some of the techniques proposed in the literature to deal with the problem of robustness to noise and tilt in the spectrum. These techniques were adapted to our present system and modified when necessary. Several approaches have been tried in conjunction with our census database and the SPHINX system. Specifically, we will describe the use of multi-style training in the case of multiple acoustical environments, and combinations of spectral subtraction and equalization algorithms.

3.1. Multi-Style Training

Multi-style training is a technique used in pattern recognition problems to increase robustness with respect to some variability or *styles*. Instead of using a model for that variability, the approach consists of including in the training a database that contains the variability.

This technique has been successfully used with Hidden Markov Models because of their powerful modeling abilities in different contexts. Lippmann *et al.* [53] pooled all the data from speakers speaking under different styles (fast, soft, angry, loud, Lombard[17]) to increase the robustness of the recognition system to different *speech styles*. Lee and Hon [47] showed that

---

[17]When the speaker is in an especially noisy environment, the produced speech, called *Lombard* speech, exhibits different characteristics than normal speech. This effect can also be simulated by recording speech from talkers in a quiet environment while they are listening to high-level noise presented through headphones.

by including speech from a large number of different speakers in a training set, speaker independence can be achieved.

An experiment was carried out in which the system was trained on all the speech recorded from both the CLSTK and the CRPZM microphones (see Table 3-1). The multi-style training is using twice as many training utterances, although they are not independent since they were recorded stereophonically. As expected, robustness is gained by using multi-style training but at the expense of sacrificing performance with respect to the case of train and test on the same conditions.

	Test **CLSTK**	Test **CRPZM**
Train **CLSTK**	85.3%	18.6%
Train **CRPZM**	36.9%	76.5%
Multi-Style	78.5%	67.9%

Table 3-1: Comparison of recognition accuracy of the SPHINX under different training conditions: close-talking microphone, Crown PZM microphone and multi-style training. Testing is done for the two microphones.

By analyzing these results it is evident that greater robustness has been obtained by the multi-style training procedure. That is, the difference in performance between the two microphones is considerably smaller than it was under normal training conditions. Unfortunately we also see that we have to pay the price of having to tolerate lower recognition rates than the those in which training and testing are done with the same microphone.

These results were not surprising as it has already been shown that although multi-style training increases the robustness of the recognition system, the accuracy is lower than that obtained with training and testing on the same condition (speech style, speaker). Robustness to different speech styles is achieved at the expense of allowing some degradation in performance over the case of training and testing on the same conditions (Lippmann *et al.* [53]). Speaker independent systems typically exhibit an error rate that is 3 or 4 times as large as speaker dependent systems given that the same amount of training data is available in both cases (Pallett *et al.* [66])

for a given system.

To achieve *microphone-independent* recognition with multi-style training, recordings from more microphones and acoustical environments will be necessary. Lee and Hon [47] used 80 speakers to achieve *speaker independence* and it is not clear how many different acoustical environments would be necessary to obtain *microphone independence*. We can expect that as recordings from more microphones are used in training, a greater degradation in performance should be observed over the case of training and testing on the same acoustical environment.

Although multi-style training is a possible solution to microphone robustness, it would be desirable to attain the robustness without the penalty of a lower overall performance. We believe that any solution to a problem of robustness will almost always imply obtaining a good model for the observed degradations.

3.2. Channel Equalization

Spectral equalization is a filtering operation used to compensate for spectral tilt. Several researchers have used some kind of spectral equalization when dealing with particular problems. It has been applied in the following contexts:

- **Speaker Verification**. Atal [3] was the first author to propose some kind of long-term normalization for a speaker verification system. The average of the cepstrum vector throughout the utterance was subtracted from each individual vector in an attempt to reduce the possibility that an imposter could be misclassified. Furui [24] confirmed in another speaker-recognition experiment that subtracting a long-term average maintained a good recognition rate while providing robustness against channels with different frequency responses. Li and Porter [49] also did some kind of speaker normalization for speaker verification.
- **Speech Styles**. Chen [8] has shown recently that the addition of a *fixed* mean vector to the cepstrum reduced the effects of different speech styles (soft, fast, shout, normal, Lombard).
- **Dereverberation**. Deconvolution in the cepstral domain has been used by Stockham *et al.* [79] to reduce the spectral tilt

effect of low-quality recordings by Caruso caused by reverberation and resonances of the recording equipment.

- **Differences in microphone and recording conditions.** Morii and Stern [57] compensated for different spectral means between recordings done with a close-talking microphone and a desk-top one in speech recognition experiments by equalizing with a fixed transfer function[18]. They observed a significant improvement in performance, although having to return to the time domain after each operation made the strategy rather inefficient.

All these approaches have in common the removal of a mean spectral vector and in all cases that mean vector was the sample mean. While some success has been achieved in specific domains of application, little integrating effort has been done to combine different sources of variability into environment independent speech recognition.

3.3. Noise Suppression by Spectral Subtraction

Spectral subtraction refers to a family of techniques designed to suppress or reduce the noise in a signal. Although it was originally proposed in the context of speech enhancement (Boll [6], Berouti *et al.* [4]), there has been a great deal of recent interest in its application to robust speech recognition. In the latter case, the end user of the processed speech is not a human being but a computer.

We present in this section an introduction to spectral subtraction with its use in speech enhancement and recognition. Then a framework for processing in the log-spectrum is presented.

3.3.1. Spectral Subtraction for Speech Enhancement

Spectral subtraction is a family of techniques that attempt to subtract the noise energy from the noisy speech energy at every frequency band. It can be formalized mathematically by assuming that the speech signal $x[m]$ is

[18]The transfer function of the equalizer was simply the ratio between the average power spectral density functions obtained for the two sets of recordings. The filtering was carried out by a method similar to the OLA (overlap and add) method.

corrupted by additive noise $n[m]$ that is uncorrelated with $x[m]$[19]:

$$y[m] = x[m] + n[m] \tag{3.1}$$

Let's define $Y(\omega)$, $X(\omega)$, and $N(\omega)$ as the power spectral densities (PSD) of the signals $y[m]$, $x[m]$ and $n[m]$ respectively. Since the signal and the noise are uncorrelated with each other, the following relationship holds for every frequency band ω_k:

$$Y(\omega_k) = X(\omega_k) + N(\omega_k) \tag{3.2}$$

If we obtain an estimate of the PSD of the corrupted signal $\hat{Y}(\omega)$ at a certain frame, and an estimate of the PSD of the noise $\hat{N}(\omega)$ from regions where no speech is present, we could obtain an estimate of the PSD of the desired signal as

$$\hat{X}(\omega_k) = \hat{Y}(\omega_k) - \hat{N}(\omega_k) \tag{3.3}$$

and thus the name of spectral subtraction. Since this estimate $\hat{X}(\omega_k)$ can go negative, Equation (3.3) is modified to disallow it by means of a half-wave rectification:

$$\hat{X}(\omega_k) = \max\{\hat{Y}(\omega_k) - \hat{N}(\omega_k), 0\} \tag{3.4}$$

We now show that under the assumption that the power spectrum is normally distributed, the ML estimate for the undegraded signal yields (3.4), the basic spectral subtraction rule. We can obtain an expression for the *joint pdf* as a product of the individual *pdfs*:

$$\begin{aligned} &p(\hat{Y}(\omega_k), \hat{N}(\omega_k) \mid Y(\omega_k), N(\omega_k)) \\ &= p(\hat{Y}(\omega_k) \mid Y(\omega_k), N(\omega_k), \hat{N}(\omega_k))\, p(\hat{N}(\omega_k) \mid Y(\omega_k), N(\omega_k)) \\ &= p(\hat{Y}(\omega_k) \mid Y(\omega_k))\, p(\hat{N}(\omega_k) \mid N(\omega_k)) \end{aligned} \tag{3.5}$$

where we have used the facts that the estimate of the corrupted signal depends only on the corrupted signal itself, and not on the noise nor its statistics. Likewise, the estimate of the noise is assumed to be independent of the corrupted signal (by doing the estimation when no signal is present for

---

[19]All notational conventions are summarized in Appendix I.

instance). By using the Gaussian assumption, the likelihood will be expressed as

$$l(\hat{Y}(\omega_k), \hat{N}(\omega_k) \mid Y(\omega_k), N(\omega_k)) = -\ln\{2\pi\sigma_y\sigma_n\} - \frac{(\hat{Y}(\omega_k) - Y(\omega_k))^2}{2\sigma_y^2} - \frac{(\hat{N}(\omega_k) - N(\omega_k))^2}{2\sigma_n^2} \tag{3.6}$$

whose maximization clearly yields $\hat{Y}(\omega_k) = Y(\omega_k)$, $\hat{N}(\omega_k) = N(\omega_k)$ if $Y(\omega_k) > N(\omega_k)$ and hence $\hat{X}(\omega_k) = \hat{Y}(\omega_k) - \hat{N}(\omega_k)$.

The idea of spectral subtraction was originally proposed by Boll [6] in the context of speech enhancement. The spectrum of the restored signal had the same phase as that of the corrupted signal, and a magnitude that was set equal to the difference between the magnitude spectra of $y[m]$ and $n[m]$. All spectra were estimated via DFTs. Boll used overlapping Hanning[20] windows and the overlap-and-add method to obtain the restored signal. His approach differs from the basic concept described above in that magnitude subtraction rather than power subtraction was used. Since speech recognition systems don't use the phase information, we will not be concerned with that issue here. The noise spectrum was estimated by averaging the magnitude of several noise frames. At low SNR the processed speech exhibited a residual noise, characterized by random spikes. He proposed additional residual noise reduction schemes, especially during non-speech activity.

Berouti *et al.* [4] proposed a modified version of the power subtraction rule (3.4) in which the amount of noise subtraction depended on the SNR of the particular frame. The oversubtraction was done to combat the residual noise that Berouti called *musical noise*.The enhanced speech obtained by these methods exhibited a greater SNR, although this processing didn't increase the intelligibility.

[20]The *Hanning* window is a raised cosine $h[m] = 0.5 - 0.5\cos(2\pi m/(N-1))$ for $0 \le m \le N-1$ and 0 otherwise.

3.3.2. Noise Subtraction in Speech Recognition

The recent use of spectral subtraction in ASR has proved more successful than in the arena of speech enhancement. Morii [57] showed in pilot experiments with SPHINX that the accuracy of the system did not deteriorate as rapidly if the waveform was enhanced by spectral subtraction techniques. Since most speech recognition systems use some kind of spectral representation as the feature space, it is unnecessary to recreate the speech signal, and one should do the subtraction directly in the transformed domain.

Porter and Boll [70] proposed MMSE estimators of DFT bins for use in speech recognition. In their work, the authors proposed a statistical characterization by assuming that the real and imaginary parts of the Fourier transform are independent and normally distributed. Under these assumptions it can be shown that the magnitude of the spectrum exhibits a Rayleigh distribution. An MMSE estimator is defined for a compression function of the magnitude spectrum as the non-linearity that minimizes:

$$E\{[f_c(\hat{X}(\omega_k)) - f_c(X(\omega_k))]^2 \mid Y(\omega_k), N(\omega_k)\} \tag{3.7}$$

whose solution $f_c(\hat{X}(\omega_k))$ equals

$$\frac{\int f_c(X(\omega_k)) \exp(-X(\omega_k)/N(\omega_k))\, I_0(2\sqrt{Y(\omega_k)X(\omega_k)}/N(\omega_k))\; p(X(\omega_k))\, dX(\omega_k)}{\int \exp(-X(\omega_k)/N(\omega_k))\, I_0(2\sqrt{Y(\omega_k)X(\omega_k)}/N(\omega_k))\; p(X(\omega_k))\, dX(\omega_k)} \tag{3.8}$$

where $I_0(x)$ is the zeroth order Bessel function. Instead of assuming a model for the distribution of speech $p(X(\omega_k))$, the integrals were approximated by a sum for all the samples of speech within a specified database. Porter and Boll tried several compression functions $f_c()$, and among them the logarithm was the one that provided the highest recognition accuracy. However, even after processing the error rate was still 5 times the error rate for clean speech.

Ephraim and Malah [15] used the same Equation (3.8) but since they assumed a Gaussian *a priori* density for speech, they were able to get a closed-form expression for (3.8), although quite complicated. Ephraim and Malah [16] also claimed that greater enhancement would be obtained if the logarithm was used as a compression function.

3.3.3. Spectral Subtraction in the Logarithm Domain

It is useful to work in the logarithm domain because distortion measures that operate in the log-spectrum have been shown to work better than the ones that operate with the regular spectrum (Gray *et al.* [31]). Boll *et al.* [7] applied the MMSE estimator he derived for DFT bins in [70] to the log-amplitudes of the outputs of a *mel-scale* filterbank[21] and obtained a greater degree of robustness. The differences with his previous approach (Porter and Boll [70]) were that the logarithm of the spectrum was used instead of the magnitude, and that filterbank outputs rather than DFT coefficients were used. The difference in recognition accuracy between the MMSE technique and Boll's own spectral subtraction technique was not as dramatic as it was when the magnitude of DFT bins were used. Two things can be learned from this experience:

- Filterbank outputs are preferred to DFT coefficients since they provide better performance. Although using different databases, the work by Boll [6] [7] indicates that the smoothing effect across frequency provided by the filterbank is very beneficial in reducing the variance of the estimate.
- Measures in the logarithm domain are more robust for recognition and they are also more consistent with the processing done in the VQ stage of SPHINX.

These results, together with the fact that a large number of speech recognition systems choose some kind of log-spectral representation, led us to investigate how the spectral subtraction rule would be in the logarithm domain.

The logarithms of the power spectral densities of the noisy speech $Y(\omega_k)$, the clean speech $X(\omega_k)$ and the noise $N(\omega_k)$ are represented by $\mathbf{Y}(\omega_k)$, $\mathbf{X}(\omega_k)$ and $\mathbf{N}(\omega_k)$ respectively. Under this notation the relationship $Y(\omega_k) = X(\omega_k) + N(\omega_k)$ can be expressed as

[21] A mel-scale filterbank has the center frequencies of the filters spaced non-linearly approximating the peripheral auditory system.

$$
\begin{aligned}
\mathbf{Y}(\omega_k) &= \ln Y(\omega_k) = \ln(X(\omega_k) + N(\omega_k)) \\
&= \ln N(\omega_k) + \ln\left(\frac{X(\omega_k)}{N(\omega_k)} + 1\right) \\
&= \mathbf{N}(\omega_k) + \ln(\exp(\mathbf{X}(\omega_k) - \mathbf{N}(\omega_k)) + 1)
\end{aligned} \tag{3.9}
$$

Let us further define the normalized values $\overline{\mathbf{Y}}(\omega_k) = \mathbf{Y}(\omega_k) - \mathbf{N}(\omega_k)$ and $\overline{\mathbf{X}}(\omega_k) = \mathbf{X}(\omega_k) - \mathbf{N}(\omega_k)$, that have the interpretation of channel SNR. Now the power addition in the logarithm domain is given by $\overline{\mathbf{Y}}(\omega_k) = \ln(\exp(\overline{\mathbf{X}}(\omega_k)) + 1)$. Similarly we can derive the power subtraction in the logarithm domain as $\overline{\mathbf{X}}(\omega_k) = \ln(\exp(\overline{\mathbf{Y}}(\omega_k)) - 1)$. Of course this expression is only defined for $\overline{\mathbf{Y}}(\omega_k) > 0$ or equivalently $Y(\omega_k) > N(\omega_k)$. The power spectral subtraction rule (3.4) can be translated into the logarithm domain as $\overline{\mathbf{X}}(\omega_k) = \max\{\ln(\exp(\overline{\mathbf{Y}}(\omega_k)) - 1), -\infty\}$.

It is clear by examining the spectral subtraction rule in the logarithm domain that making $\overline{\mathbf{X}}(\omega_k) = -\infty$ will yield an infinite value for VQ distortions, that will in turn result in errors in the recognizer. The $-\infty$ appears because we have implicitly assumed that the original signal was completely noise-free, which is not realistic. Since our clean speech will still have some noise (quantization, aliasing, estimation error, etc) the SNR will be high but not infinite. Our goal is to increase the SNR of the corrupted signal to the level it was before.

If this *a priori* information is taken into account, the spectral subtraction rule in the logarithm domain will have the form

$$
\overline{\mathbf{X}}(\omega_k) = \max\{\ln(\exp(\overline{\mathbf{Y}}(\omega_k)) - 1), \mathbf{X}_{th}(\omega_k) - \hat{\mathbf{N}}(\omega_k)\} \tag{3.10}
$$

where $\hat{\mathbf{N}}(\omega_k)$ is the estimate of the noise level and $\mathbf{X}_{th}(\omega_k)$ is the signal floor level. This new spectral subtraction will take the input log-power $\mathbf{Y}(\omega_k)$ and produce an estimate of the log-power of the clean signal $\mathbf{X}(\omega_k)$ given by $\hat{\mathbf{X}}(\omega_k) = \max\{\mathbf{X}_{th}(\omega_k), \ln(\exp(\hat{\mathbf{Y}}(\omega_k)) - \exp(\hat{\mathbf{N}}(\omega_k)))\}$. Figure 3-1 shows a typical curve that can be implemented as a table lookup. Van Compernolle

[11] also proposed the same transformation.

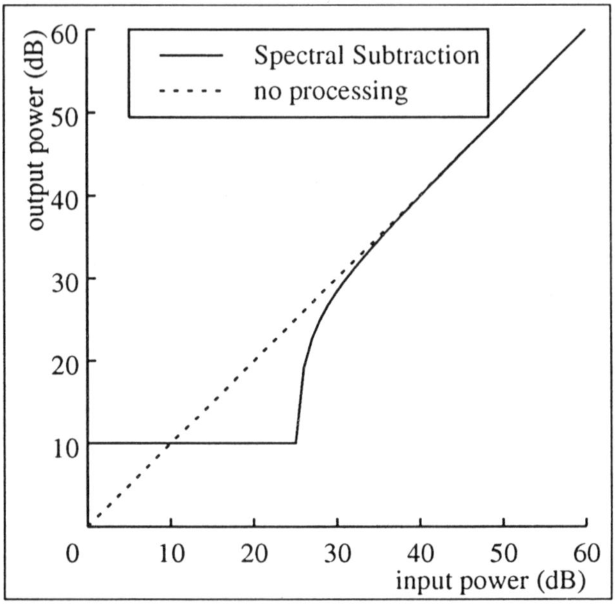

Figure 3-1: Spectral subtraction curve for $\hat{\mathbf{N}}(\omega_k) = 25\,\text{dB}$ and $\mathbf{X}_{th}(\omega_k) = 10\,\text{dB}$.

3.4. Experiments with Sphinx

In this section we describe a series of experiments performed using the census database, running the SPHINX system with a number of spectral equalization and noise suppression algorithms. In spectral subtraction and equalization it is assumed that the speech signal $x[m]$ is degraded by linear filtering and/or uncorrelated additive noise, as depicted in Figure 3-2. The goal of the compensation is to reverse the effects of these degradations.

Spectral equalization techniques attempt to compensate for the filter $h[m]$, while spectral subtraction techniques attempt to remove the effects of the noise from the signal. We compare the performance of several different implementations of spectral subtraction and equalization techniques in Table 3-7, which we refer as EQUAL, PSUB, MMSE1 and MSUB. These

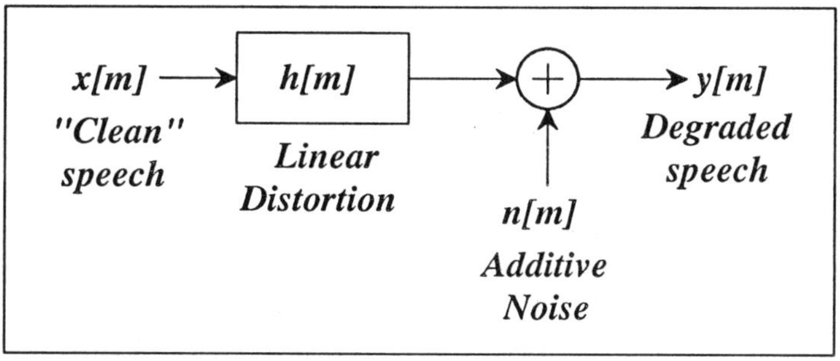

Figure 3-2: Model of the degradation.

algorithms were applied *only* to the CRPZM speech, as the goal of the compensation schemes is to be able to use the HMM models trained with the CLSTK. In this section we describe these algorithms and examine the extent to which they make the SPHINX system more environmentally robust.

To illustrate their performance, we will consider three-dimensional spectrograms of one sample utterance for all algorithms. In Figures 3-3 and 3-4 we show 3-D plots of the word *yes* recorded with the CLSTK and the CRPZM microphone respectively. These 3-D plots are a time-frequency representation of the signal obtained by taking the DFT of the cepstral vectors computed by SPHINX's front-end. The frequency axis is warped by the bilinear transform (See Chapter 7). Since VQ labels reflect different spectral shapes, the labels assigned to the noise segment of the CRPZM (up to 330 ms, and beyond 750 ms) are very different than the ones assigned for the CLSTK. This is a major cause for the errors observed in Table 3-2.

3.4.1. EQUAL Algorithm

EQUAL is a spectral equalizationalgorithm that compensates for the effects of the linear filtering, but not the additive noise, and was first described in [78] (Stern and Acero). This algorithm is inspired by work done by Stockham *et al.* [79] in blind deconvolution. Compensation for linear convolution is accomplished by multiplying the spectrum of the input signal by the spectrum of the compensating filter[22]. Therefore the cleaned log-

---

[22]Since the cepstrum is the inverse Fourier transform of the logarithm of the magnitude of the spectrum, and we are not interested in the phase for recognition, this corresponds to a simple additive correction of the cepstrum vector.

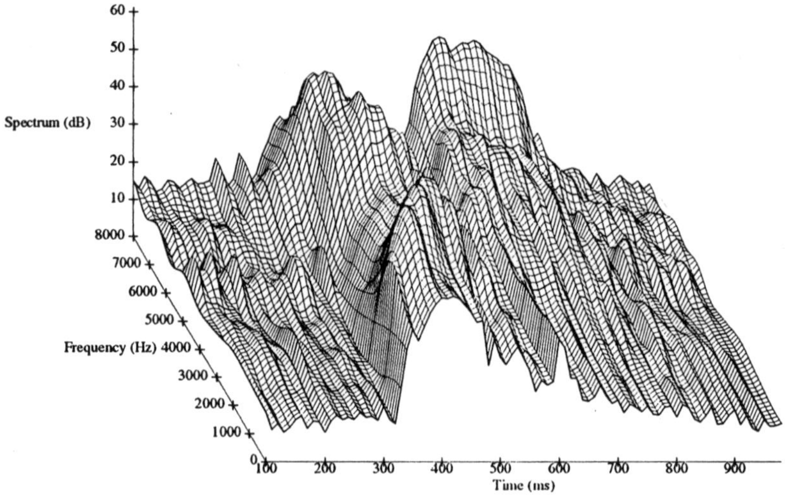

Figure 3-3: 3D Spectrogram of the utterance *yes* recorded with the CLSTK microphone with no processing.

TRAIN TEST	CLSTK CLSTK	CLSTK CRPZM	CRPZM CLSTK	CRPZM CRPZM
BASE	85.3%	18.6%	36.9%	76.5%

Table 3-2: Baseline recognition accuracy for the CLSTK and CRPZM.

spectrum $\mathbf{X}(\omega_k)$ can be estimated from the input log-spectrum $\mathbf{Y}(\omega_k)$ and the equalization $\mathbf{Q}(\omega_k)$:

$$\hat{\mathbf{X}}(\omega_k) = \mathbf{Y}(\omega_k) + \mathbf{Q}(\omega_k) \tag{3.11}$$

The equalization $\mathbf{Q}(\omega_k)$ is estimated as the difference between the average log-spectra for the two microphones, computed during speech activity (an approximate MMSE estimate if all the frames in the calculation can be considered noise free). Alternatively, the compensation can be applied as an additive correction in the cepstral domain to the CRPZM speech.

In Figure 3-5 we can see the spectrum of the sample utterance for the CRPZM after processing by this algorithm. By comparing the processed

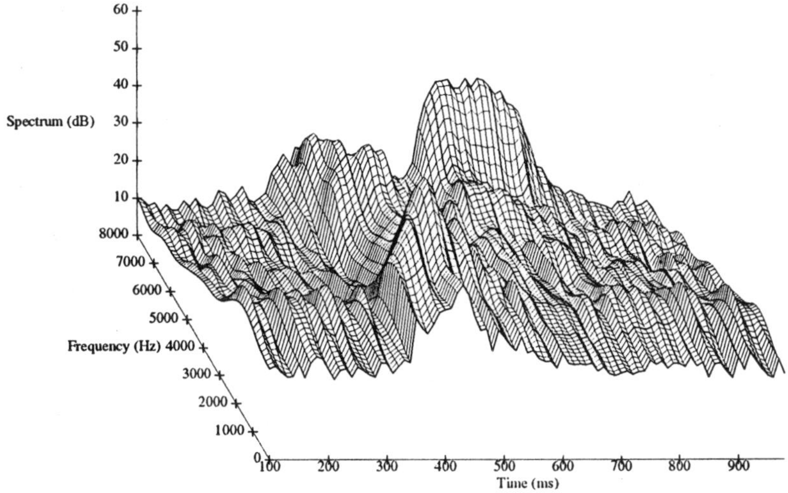

an251-fash-c

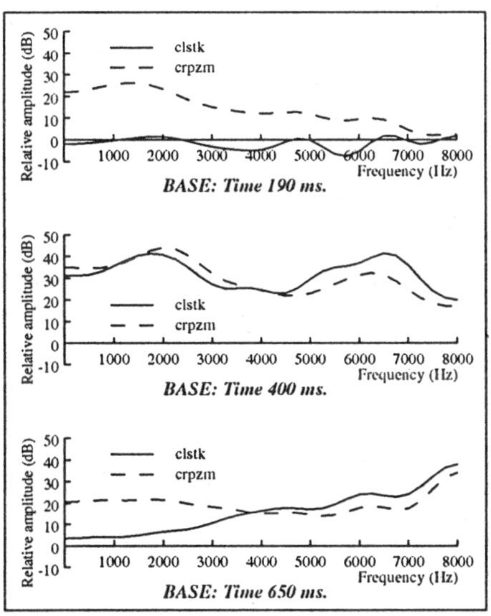

Figure 3-4: 3D Spectrogram of the utterance *yes* recorded with the CRPZM microphone with no processing. Spectra at times 190 ms (silence region), 400 ms (vowel) and 650 ms (fricative) are plotted for both the CLSTK and the CRPZM.

TRAIN TEST	CLSTK CLSTK	CLSTK CRPZM	CRPZM CLSTK	CRPZM CRPZM
BASE	85.3%	18.6%	36.9%	76.5%
EQUAL	N/A	38.3%	50.9%	76.5%

Table 3-3: Comparison of the baseline recognition accuracy and the EQUAL spectral equalization algorithm. EQUAL was only applied to the CRPZM.

frame at time 400 ms of Figure 3-5 with that of Figure 3-4, we see that some spectral tilt has been removed. There is still some residual spectral tilt, because a single equalization function is used for all the speakers and utterances, whereas a different one would be needed (The spectral tilt varies from speaker to speaker, due to changes in the relative location of the mouth and the microphone.). The results in Table 3-3, show some improvement in performance for the cross conditions over the baseline case.

3.4.2. PSUB Algorithm

PSUB is an implementation of the power spectral subtraction rule in the logarithm domain (3.10). The transformation is accomplished by converting the cepstral coefficients to log-power coefficients using a 32-point inverse DFT (for the 32 real and even cepstral coefficients), performing the subtraction in these 32 frequency bands and converting once again to the cepstral domain by another DFT.

The estimate of the noise is computed by averaging the cepstrum vectors whose total energy $\mathbf{c}_i[0]$ fell under a threshold[23]. In our case this threshold is empirically set to the minimum $\mathbf{c}_i[0]$ in the utterance plus 1.0, although the performance was not very sensitive to small variations of this value.

Finally, we note that the noise floor was picked in such a way that the

[23]In this case speech/noise discrimination was done on a frame-by-frame basis, assuming that different speech frames were statistically independent. Furthermore, since only $\mathbf{c}_i[0]$ was used for the classification, the optimum scheme is a comparison with a threshold.

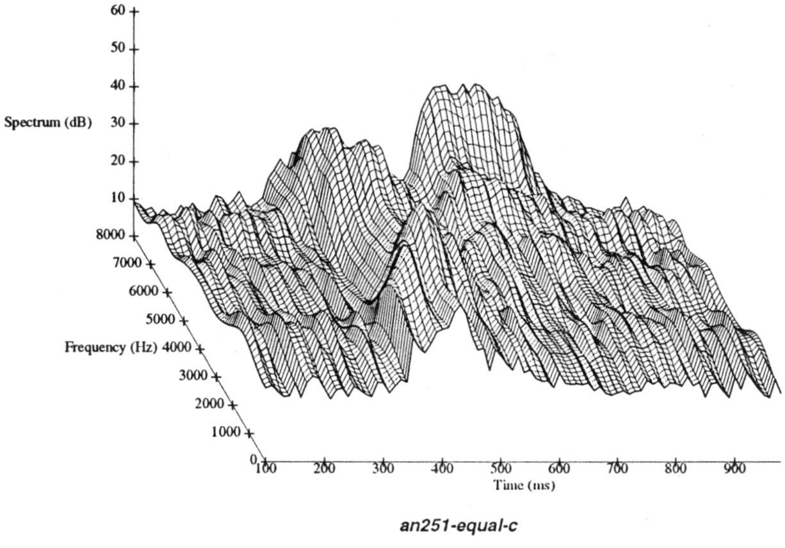

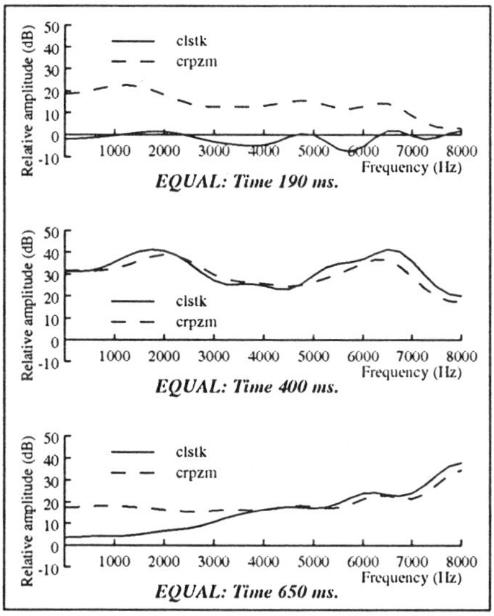

Figure 3-5: 3D Spectrogram of the utterance *yes* recorded with the CRPZM microphone with EQUAL algorithm. Spectra at times 190 ms (silence region), 400 ms (vowel) and 650 ms (fricative) are plotted for both the CLSTK and the CRPZM.

dynamic range for the utterance in **c**[0] was always 10.0 (equivalent to 44 dB). This was the optimum value as found by experimentation. We have found that the recognition accuracy was not very sensitive to this level.

TRAIN TEST	CLSTK CLSTK	CLSTK CRPZM	CRPZM CLSTK	CRPZM CRPZM
BASE	85.3%	18.6%	36.9%	76.5%
PSUB	N/A	38.6%	70.6%	70.1%

Table 3-4: Comparison of the baseline recognition accuracy with the PSUB power subtraction algorithm. PSUB was only applied to the CRPZM.

In Figure 3-6 we see the spectrum of the example utterance for the CRPZM after processing by this algorithm, including the presence at low SNR of what Berouti *et al.* [4] called *musical noise*. Although the mean has been lowered, the residual noise exhibits spikes uncharacteristic of speech frames (See frames at time 190 ms and 650 ms). Since the vector quantizer essentially clusters spectral shapes, these spikes will cause large errors in the VQ labeler that in turn will translate into errors in the recognizer. Nevertheless, the evaluation results in Table 3-4 show some additional improvement in performance over the baseline condition when the PSUB algorithm is applied to the CRPZM speech.

3.4.3. MSUB Algorithm

To ameliorate the problem of the *musical noise* present in the PSUB algorithm we developed the MSUB algorithm, first described by Stern and Acero in [78]. We introduced over-subtraction at low SNR to further suppress the *musical noise* following an idea originally suggested by Berouti *et al.* [4]. We note that we subtract the magnitudes of spectra as did Boll [6] rather than the more intuitively appealing spectral power because we found that magnitude subtraction provided greater recognition accuracy. The amount of over-subtraction or under-subtraction is a function of the instantaneous SNR according to the transformation shown in Figure 3-7. This transformation was developed empirically by analyzing 3-D spectrograms obtained with the MSUB algorithm.

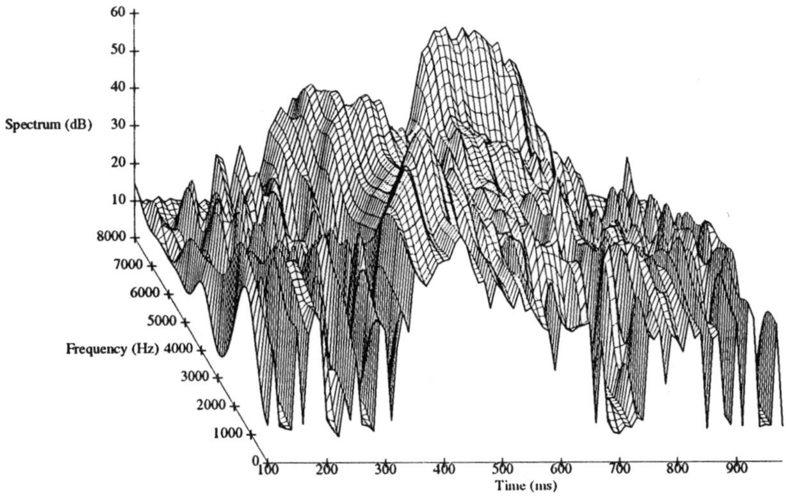

an251-psub-c

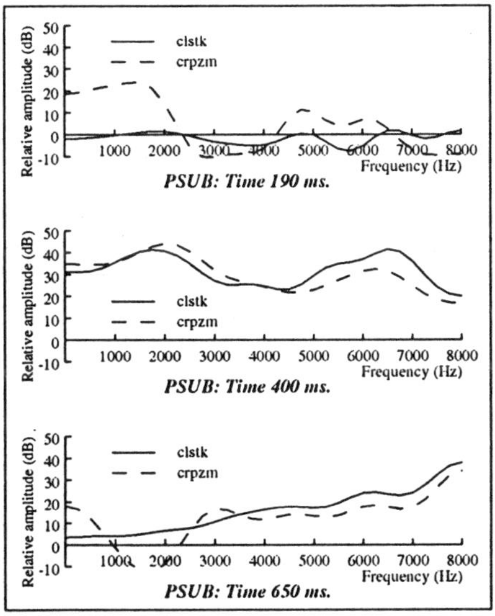

Figure 3-6: 3D Spectrogram of the utterance *yes* recorded with the CRPZM microphone with PSUB algorithm. Spectra at times 190 ms (silence region), 400 ms (vowel) and 650 ms (fricative) are plotted for both the CLSTK and the CRPZM.

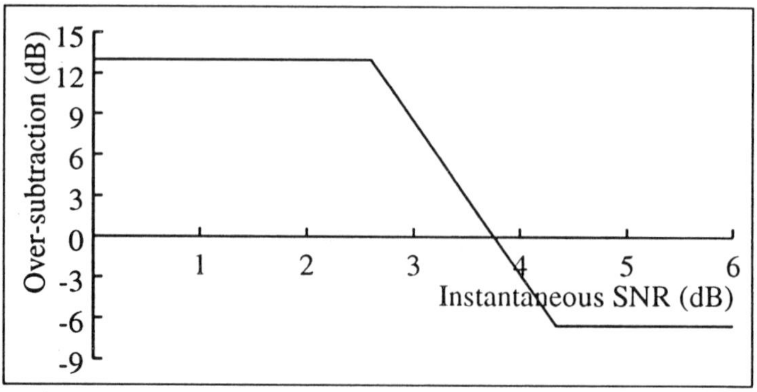

Figure 3-7: Amount of over and under subtraction used in the MSUB algorithm as a function of the instantaneous SNR.

TRAIN TEST	CLSTK CLSTK	CLSTK CRPZM	CRPZM CLSTK	CRPZM CRPZM
BASE	85.3%	18.6%	36.9%	76.5%
PSUB	N/A	38.6%	70.6%	70.1%
MSUB	N/A	63.6%	71.7%	71.3%

Table 3-5: Comparison of the baseline recognition accuracy with the PSUB power subtraction and the MSUB magnitude subtraction algorithms. PSUB and MSUB were only applied to the CRPZM.

In Figure 3-8 we show the spectrum of the example utterance for the CRPZM after processing by the MSUB algorithm. We note that MSUB suppresses the musical noise present in the PSUB algorithm (See frames at time 190 ms and 650 ms). The evaluation results in Table 3-5 show the effectiveness of using the magnitude rather than the power, and the use of the overall instantaneous SNR in determining under and over subtraction. The MSUB algorithm performs considerably better than the PSUB algorithm.

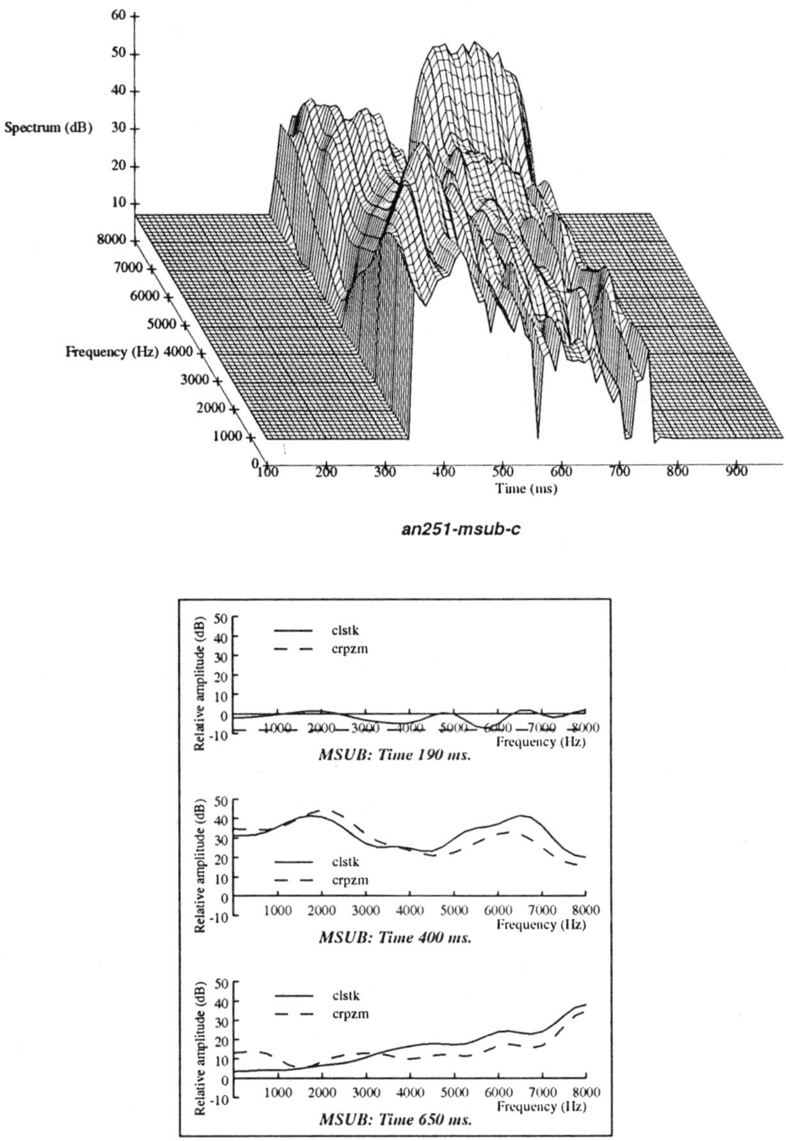

Figure 3-8: 3D Spectrogram of the utterance *yes* recorded with the CRPZM microphone with MSUB algorithm. Spectra at times 190 ms (silence region), 400 ms (vowel) and 650 ms (fricative) are plotted for both the CLSTK and the CRPZM.

3.4.4. MMSE1 Algorithm

While the vector quantizer uses an MMSE criterion, the noise suppression algorithms derived so far are not goal-directed. Perhaps overall system performance could be improved if both modules would use the same MMSE criterion. Porter and Boll [70] realized this fact and used an MMSE estimator based on Equations (3.7) and (3.8). He suggested the use of the logarithm as a compression function. We will follow a similar approach here.

Let $\mathbf{N}(\omega_k)$ be the log-energy of the noise in band k, and $\mathbf{Y}(\omega_k)$ the log-energy of the corrupted signal in band k. We define the SNR in band k or normalized input $\overline{\mathbf{Y}}(\omega_k)$ as:

$$\overline{\mathbf{Y}}(\omega_k) = \mathbf{Y}(\omega_k) - \mathbf{N}(\omega_k) \tag{3.12}$$

Then, the estimate of the log-energy of the restored signal in band k, $\hat{\mathbf{X}}(\omega_k)$, can be expressed as follows:

$$\hat{\mathbf{X}}(\omega_k) = \mathbf{Y}(\omega_k) + f(\overline{\mathbf{Y}}(\omega_k)) \tag{3.13}$$

Porter and Boll [70] used the assumption that the magnitude of the noisy speech had a Rayleigh distribution in order to derive his estimator. In our work, the transformation f was derived directly from the stereo database as the transformation that minimized:

$$E\{[\mathbf{X}(\omega_k) - \hat{\mathbf{X}}(\omega_k)]^2\} = E\{[\mathbf{X}(\omega_k) - \mathbf{Y}(\omega_k) - f(\overline{\mathbf{Y}}(\omega_k))]^2\} \tag{3.14}$$

where the $\mathbf{Y}(\omega_k)$ are taken from the CRPZM speech and $\mathbf{X}(\omega_k)$ from the CLSTK speech for the same utterance. The function f is the one that makes the CRPZM speech as similar to the CLSTK as possible in the sense of minimum mean squared error. To compute the new transformation curves, for every pair of stereo utterances from the database a gain normalization was performed so that the maximum zeroth cepstral component $\mathbf{c}[0]$ in both utterances is 0.

The expectation is replaced by a summation over all the frames in the training corpus for all 32 frequency bands. The function f was discretized into 25 intervals separated 1 dB each. The result of the optimization was the value of f for $j = 0, \ldots, 25$ in steps of $\Delta_{SNR} = 1dB$.

$$f[j] = \frac{\sum_{i=0}^{N-1} \sum_{k=0}^{k=32} [\mathbf{X}_i(\omega_k) - \mathbf{Y}_i(\omega_k)]\, \delta[\overline{\mathbf{Y}}_i(\omega_k) - j\Delta_{SNR}]}{\sum_{i=0}^{N-1} \sum_{k=0}^{k=32} \delta[\overline{\mathbf{Y}}_i(\omega_k) - j\Delta_{SNR}]} \tag{3.15}$$

where $\mathbf{X}_i(\omega_k)$ and $\mathbf{Y}_i(\omega_k)$ represent the log-energy of frame i at frequency band k for the CLSTK and CRPZM respectively. $\overline{\mathbf{Y}}_i(\omega_k)$ is the quantized SNR in frequency band k of frame i in the CRPZM, and $\delta[i]$ is the delta of Kronecker. The sum is carried out for all the N frames in the database.

The MMSE1 algorithm applies the transformation to the 32 frequency bands of the CRPZM speech that minimizes the mean squared error relative to the CLSTK speech. The calculation of the compensating transformation was simpler with the use of the stereo database than with the equations derived by Porter and Boll [70] and Ephraim and Malah [15], without suffering from modeling inaccuracies.

TRAIN TEST	CLSTK CLSTK	CLSTK CRPZM	CRPZM CLSTK	CRPZM CRPZM
BASE	85.3%	18.6%	36.9%	76.5%
PSUB	N/A	38.6%	70.6%	70.1%
MSUB	N/A	63.6%	71.7%	71.3%
MMSE1	N/A	48.7%	68.7%	71.4%

Table 3-6: Comparison of baseline recognition accuracy with the PSUB, MSUB and MMSE1 algorithms. These algorithms were only applied to the CRPZM.

In Figure 3-9 we can see the spectrum of the sample utterance for the CRPZM after processing by the MMSE1 algorithm. This algorithm performs noticeably better than the PSUB at low SNR (See frames at time 190 ms and 650 ms), although not as well as the MSUB algorithm. The evaluation results are shown in Table 3-6. The MMSE criterion outperforms the power subtraction rule as we expected. However, the MSUB algorithm still has higher accuracy despite not being goal-directed. We believe that the power of the MSUB algorithm stems from using different transformation curves for

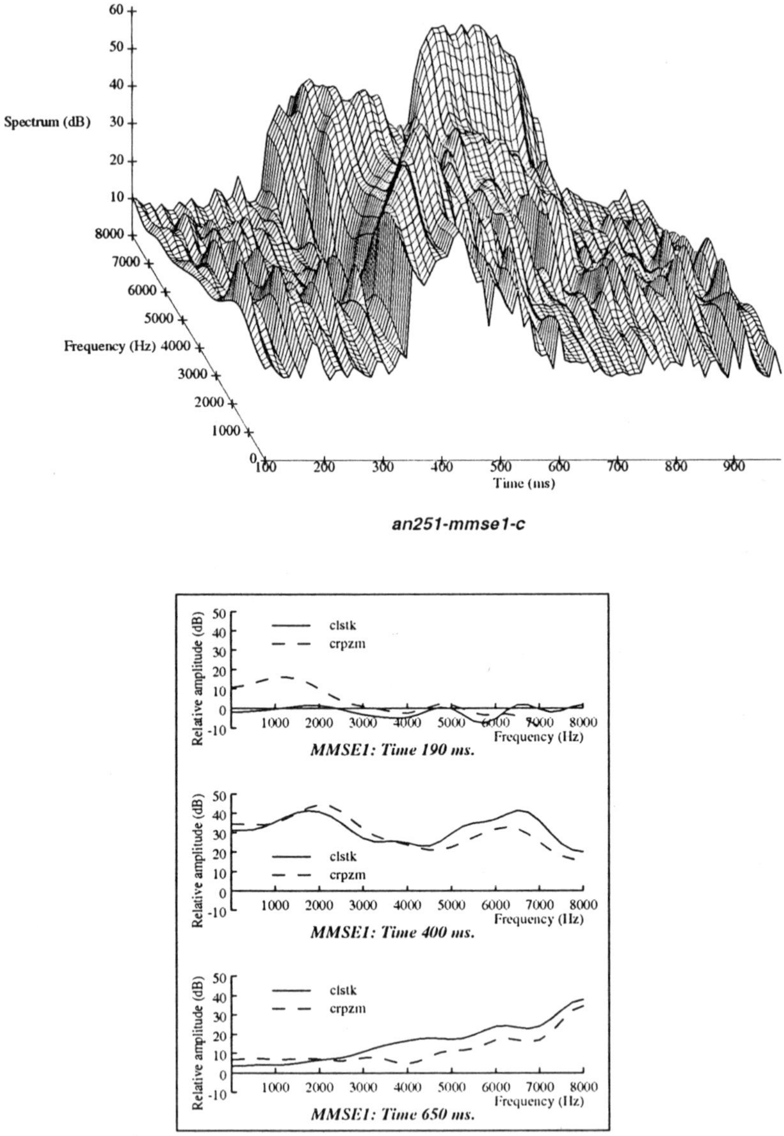

Figure 3-9: 3D Spectrogram of the utterance *yes* recorded with the CRPZM microphone with cascade of EQUAL and MMSE1 algorithms. Spectra at times 190 ms (silence region), 400 ms (vowel) and 650 ms (fricative) are plotted for both the CLSTK and the CRPZM.

every frame SNR, as opposed to one single curve.

3.4.5. Cascade of EQUAL and MSUB

We applied a cascade of our channel equalization algorithm EQUAL and our noise suppression algorithms MMSE1 and MSUB, in an attempt to obtain an increased performance, since each algorithm combats a different kind of distortion.

Since equalization is a linear operation and the noise suppression algorithms are not, we opted to first perform the EQUAL algorithm and then the noise suppression. To apply the MMSE1 algorithm, another transformation function was calculated that minimized the squared error between the CLSTK speech and the equalized CRPZM speech. Figure 3-10 shows the sample utterance after this cascade of EQUAL and MMSE1. By comparing the frame at time 400 ms in Figures 3-10 and 3-9 we see that this processing is effective in combating spectral tilt too. The recognition results in Table 3-7 show the combined benefits of EQUAL and MMSE1. Unfortunately this combination does not perform better than the MSUB algorithm that only does noise suppression.

Following the success of the cascade of EQUAL and MMSE1, we proceeded to apply the cascade of EQUAL with our best noise subtraction algorithm MSUB. Again, we first performed the equalization and then the noise subtraction. In Figure 3-11 we present the sample utterance after this combined processing. The recognition results of this cascade of EQUAL and MSUB shown in Table 3-7 are disappointing, since the combined version is not doing better than the single MSUB. We believe that this is because of the non-linear interaction of noise and spectral tilt.

3.4.6. Results and Discussion

The results of EQUAL, PSUB, MMSE1 and MSUB are shown in Table 3-7. There are several observations:

- The use of *a priori* information in MMSE1 resulted in a higher accuracy than the empirical power spectral subtraction rule in PSUB.
- The dependence on the instantaneous SNR was very beneficial,

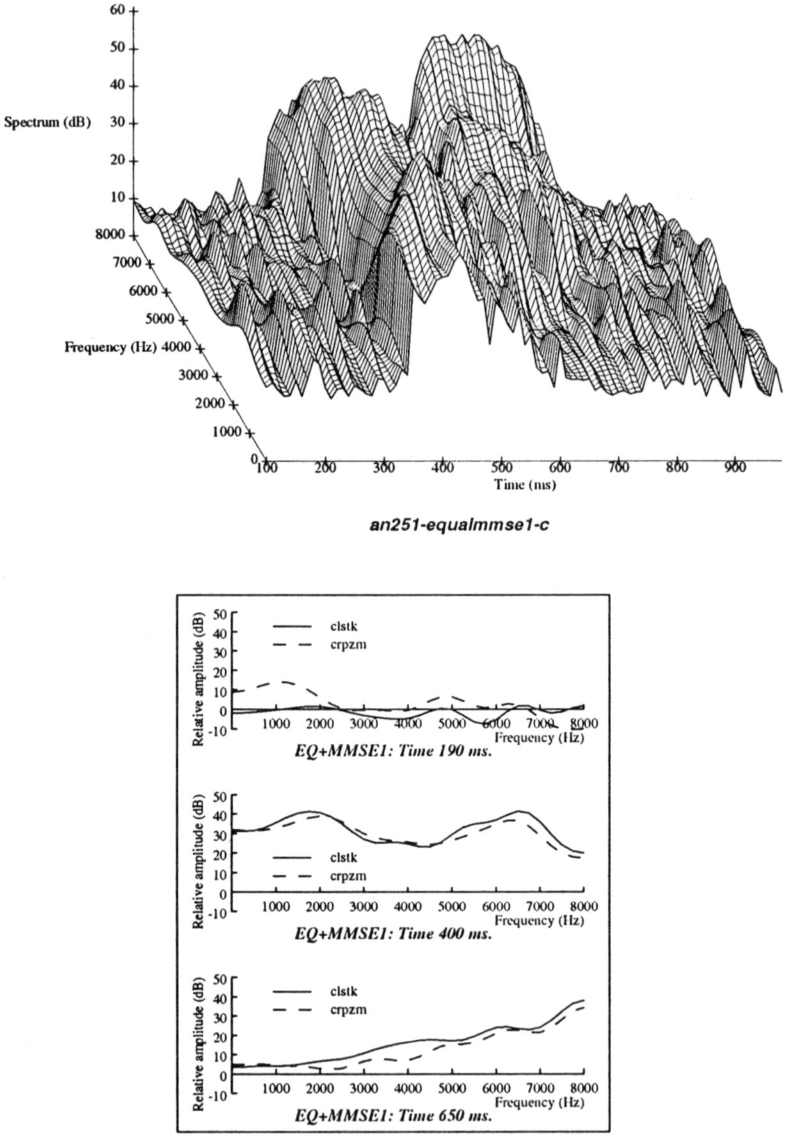

Figure 3-10: 3D Spectrogram of the utterance *yes* recorded with the CRPZM microphone with the cascade of EQUAL and MMSE1 algorithms. Spectra at times 190 ms (silence region), 400 ms (vowel) and 650 ms (fricative) are plotted for both the CLSTK and the CRPZM.

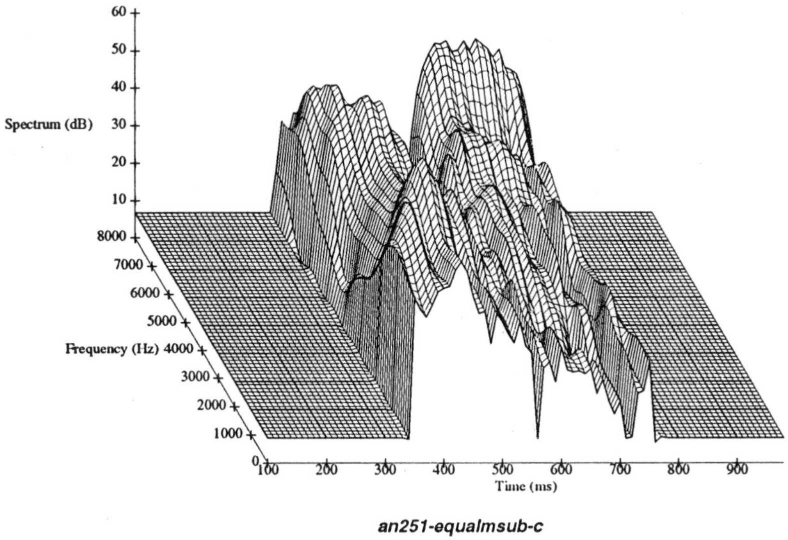

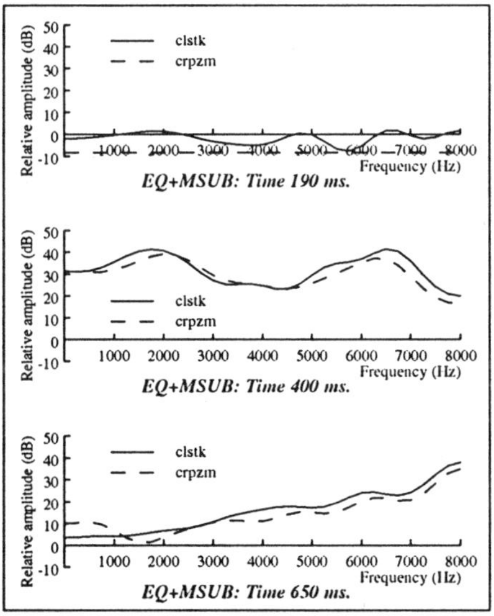

Figure 3-11: 3D Spectrogram of the utterance *yes* recorded with the CRPZM microphone with the cascade of EQUAL and MSUB algorithms. Spectra at times 190 ms (silence region), 400 ms (vowel) and 650 ms (fricative) are plotted for both the CLSTK and the CRPZM.

as noted by the difference in performance between PSUB and MSUB.

- The cascade of spectral equalization EQUAL and noise suppression MMSE1 performed considerably better than either of those alone. Unfortunately, this cascade did not yield any further improvement in recognition accuracy over the MSUB algorithm that only accomplishes noise suppression. Furthermore, a cascade of EQUAL and MSUB did not increase the accuracy over the MSUB algorithm alone. We conjecture that since noise is an additive distortion and channel equalization is a multiplicative distortion in the frequency domain, the two effects interact nonlinearly.

TRAIN TEST	CLSTK CLSTK	CLSTK CRPZM	CRPZM CLSTK	CRPZM CRPZM
BASE	85.3%	18.6%	36.9%	76.5%
EQUAL	N/A	38.3%	50.9%	76.5%
PSUB	N/A	38.6%	70.6%	70.1%
MSUB	N/A	63.6%	71.7%	71.3%
MMSE1	N/A	48.7%	68.7%	71.4%
EQ+MMSE1	N/A	61.4%	75.8%	74.3%
EQ+MSUB	N/A	62.1%	73.7%	71.4%

Table 3-7: Performance of different equalization and spectral subtraction algorithms. All the algorithms were applied only to the CRPZM.

As a way of unifying all the approaches, we can view the different subtraction algorithms as different transformation curves that relate the effective SNR of the input and output for every frequency channel. We note that the MSUB algorithm is not represented by a single curve by but a family of curves that depended on the instantaneous SNR of the frame. Some of these curves are shown in Figure 3-12.

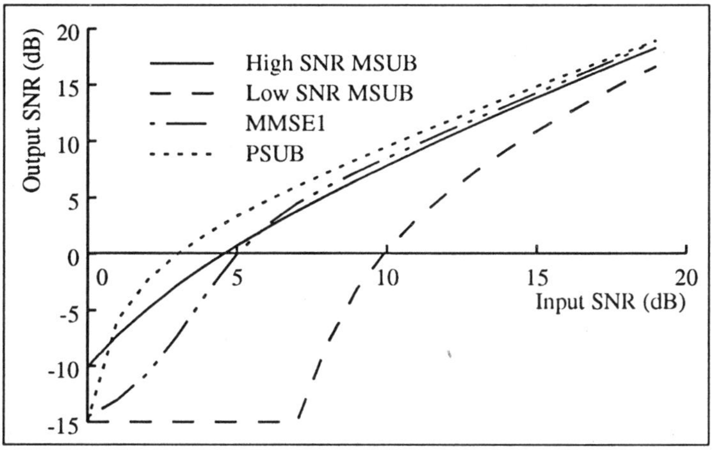

Figure 3-12: Input-Output transformation curves for PSUB, MSUB and MMSE1. The channel SNR is defined as the log-power of the signal in a frequency band minus the log-power of the noise in that band. The transformation for MSUB is not a single curve but a family of curves that depend on the total SNR for a given frame.

3.5. Summary

In this chapter we have presented several algorithms to increase the robustness of the system for the "cross conditions". Multi-Style training is a plausible alternative to the problem of robustness although at the expense of a lower recognition accuracy than when training and testing are done on the same environment.

The algorithm EQUAL applied a spectral equalization to the speech recorded with the CRPZM microphone to compensate for differences in transfer functions between training and test corpora.

A number of noise suppression techniques were discussed in this chapter. First, we described in detail the power spectral subtraction rule. An algorithm called PSUB directly implemented this rule in the logarithm domain. The MSUB algorithm was developed to combat the *musical noise* characteristic of standard spectral subtraction. Magnitude subtraction was used in MSUB as opposed to power subtraction because it provided a higher accuracy. The need for a uniform criterion in the front-end led us to try noise suppression techniques based on an MMSE criterion.

Cascading of equalization and spectral subtraction combined the improvements of both techniques although the recognition accuracy was still worse than for the MSUB algorithm. A cascade of EQUAL and MSUB did not perform better presumably because of the non-linear interaction of the equalization and noise subtraction.

For the most part these algorithms provide increasing degrees of compensation, but their recognition accuracy under the "cross" conditions is still worse than that obtained with the system is trained and tested on the CRPZM. The problem of the nonlinear interaction of the subtraction and normalization processes motivated us to consider new algorithms which jointly compensate for noise and filtering. We discuss several such algorithms in the next chapter.

4 The SDCN Algorithm

In this chapter we introduce the concept of joint normalization of noise and spectral tilt as well as the idea of preprocessing in the cepstral domain as opposed to the frequency domain. We first define the model of the environment that will be used in the remaining of the monograph and we introduce techniques that accomplish the processing in the cepstral domain.

We develop the MMSEN algorithm as an extension of the MMSE1 algorithm of Chapter 3. The MMSEN algorithm is able to cope with colored noise and spectral tilt jointly, and yields higher accuracy than a cascade of equalization and MMSE1.

By looking at the cepstral domain instead of the frequency domain we introduce the SDCN algorithm, that performs a cepstral compensation that depends on the SNR of the input frame exclusively. This algorithm is simple and effective.

4.1. A Model of the Environment

In this section we introduce the model of the environment as well as some notation conventions that will be used in the remaining of this monograph. We will present relationships in the cepstral domain and define the concept of correction vectors.

The model of Figure 4-1 describes the two kinds of degradation that we have seen in the previous chapter: additive noise and linear filtering. We further assumed that the noise, which is colored[24] in general, was independent of the signal. Although we could have chosen to add the noise first and filter later, the model of Figure 4-1 introduces no loss of generality and has the advantage that observation of $y[m]$ when no signal is present provides us the noise directly and not filtered noise. For this work we will assume that the filter $h[m]$ is *time-invariant* and that the noise is *stationary* so that the parameters of the model are assumed to be fixed during the course of our observation.

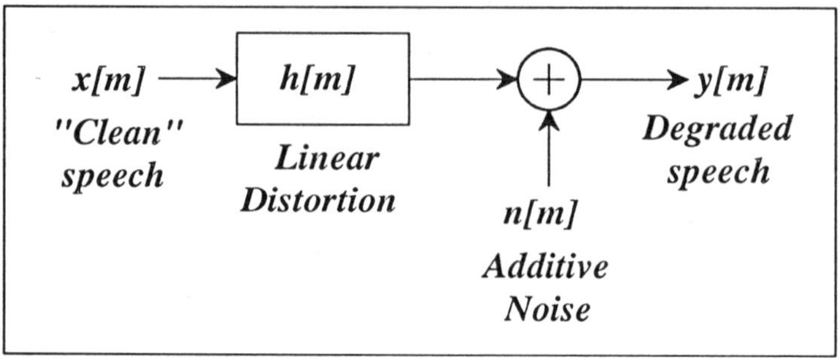

Figure 4-1: Model of the degradation.

If the input to the linear time-invariant filter $h[m]$ is a stationary random process $x[m]$ with Power Spectral Density (PSD) $X(\omega)$, we know [67] that the PSD of the output process $v[m]$ is

$$V(\omega) = X(\omega)\,|H(\omega)|^2 \tag{4.1}$$

Furthermore, if a stationary random process $n[m]$ that is uncorrelated with $v[m]$ is added to it, the PSD of the resulting random process $y[m]$ is

$$Y(\omega) = V(\omega) + N(\omega) \tag{4.2}$$

Combining (4.1) and (4.2) we obtain an expression that relates the PSD of the input, noise, output and the transfer function of the filter:

$$Y(\omega) = X(\omega)\,|H(\omega)|^2 + N(\omega) \tag{4.3}$$

Let's express Equation (4.3) in the cepstral domain rather than in the

---

[24] *White* noise presents an even distribution of power for all frequencies, whereas *colored* noise has more power in some frequency bands than others.

frequency domain. To do that, we need to define the following cepstrum vectors **x**, **n**, **y** and **q** as the following Inverse Discrete Fourier Transforms (IDFT):

$$\mathbf{x} = \text{IDFT}\ \{ \ln X(\omega_k) \} \tag{4.4}$$
$$\mathbf{n} = \text{IDFT}\ \{ \ln N(\omega_k) \} \tag{4.5}$$
$$\mathbf{y} = \text{IDFT}\ \{ \ln Y(\omega_k) \} \tag{4.6}$$
$$\mathbf{q} = \text{IDFT}\ \{ \ln |H(\omega_k)|^2 \} \tag{4.7}$$

Taking natural logarithms and IDFT on (4.3) and after some algebraic manipulation using (4.4)-(4.7), we obtain:

$$\mathbf{y} = \mathbf{x} + \mathbf{q} + \mathbf{r}(\mathbf{x},\mathbf{n},\mathbf{q}) \tag{4.8}$$

or alternatively

$$\mathbf{y} = \mathbf{n} + \mathbf{s}(\mathbf{x},\mathbf{n},\mathbf{q}) \tag{4.9}$$

where the correction vectors $\mathbf{r}(\mathbf{x},\mathbf{n},\mathbf{q})$ and $\mathbf{s}(\mathbf{x},\mathbf{n},\mathbf{q})$ are given by

$$\mathbf{r}(\mathbf{x},\mathbf{n},\mathbf{q}) = \text{IDFT}\ \{\ln\ (1 + e^{\text{DFT}\ [\mathbf{n} - \mathbf{q} - \mathbf{x}]})\} \tag{4.10}$$

$$\mathbf{s}(\mathbf{x},\mathbf{n},\mathbf{q}) = \text{IDFT}\ \{\ln\ (1 + e^{\text{DFT}\ [\mathbf{x} + \mathbf{q} - \mathbf{n}]})\} \tag{4.11}$$

where (4.8) and (4.9) can be used interchangeably. The correction vectors **r** and **s** are uniquely related given the knowledge of **n**, **q** and **x**. In this monograph we will mainly use **r** and the expression (4.8).

Although Equations (4.8) and (4.9) involve only vector additions, the presence of the correction vectors of Equations (4.10) and (4.11) may at first indicate that the relationship was simpler in the frequency domain than in the cepstral domain. However, these vectors could be precomputed as we will see in Chapters 5 and 6.

In summary, the model in Figure 4-1 can be characterized in the frequency domain by (4.3), or alternatively in the cepstral domain by (4.8) and (4.9) with the newly introduced correction vectors in (4.10) and (4.11).

4.2. Processing in the Frequency Domain: The MMSEN Algorithm

Although the approaches developed in Chapter 3 were relatively successful, they present some problems. The amount of noise subtraction given by the transformation curves of Figure 3-12 is the same for all frequencies, which is unable to deal with colored noise as in our case. To cope with these problems, we propose the MMSEN algorithm.

The MMSEN algorithm applies an *independent* transformation for every frequency band that minimizes the mean squared error between the CLSTK and the CRPZM speech. Equation (3.12) is still valid and Equation (3.13) becomes

$$\hat{\mathbf{X}}(\omega_k) = \mathbf{Y}(\omega_k) + f_k(\overline{\mathbf{Y}}(\omega_k)) \tag{4.12}$$

where there are 32 independent f_k transformations that are chosen to minimize

$$E\{[\mathbf{X}(\omega_k) - \hat{\mathbf{X}}(\omega_k)]^2\} = E\{[\mathbf{X}(\omega_k) - \mathbf{Y}(\omega_k) - f_k(\overline{\mathbf{Y}}(\omega_k))]^2\} \tag{4.13}$$

Minimization of Equation (4.13) yields

$$f_k[j] = \frac{\sum_{i=0}^{N-1} [\mathbf{X}_i(\omega_k) - \mathbf{Y}_i(\omega_k)]\,\delta[\overline{\mathbf{Y}}_i(\omega_k) - j\Delta_{SNR}]}{\sum_{i=0}^{N-1} \delta[\overline{\mathbf{Y}}_i(\omega_k) - j\Delta_{SNR}]} \tag{4.14}$$

where $\mathbf{X}_i(\omega_k)$ and $\mathbf{Y}_i(\omega_k)$ represent the log-energy of frame i at frequency band k for the CLSTK and CRPZM respectively. $\overline{\mathbf{Y}}_i(\omega_k)$ is the SNR in frequency band k of frame i in the CRPZM. The sum is carried out for all the N frames in the database.

To compute the new transformation curves, for every pair of stereo utterances from the database a gain normalization was performed so that the maximum $\mathbf{c}_{max}[0]$[25] in both utterances is 0. For every frequency band the difference between the CLSTK and the CRPZM was computed as a function of the channel SNR. Figure 4-2 shows the transformation curves for frequency bands of 0, 1, 2, 3, 4, 5, 6, and 8 kHz and its comparison with the MMSE curve averaged over all frequencies.

We can see that at low frequencies (below 4 kHz), the amount of noise subtraction is greater than at high frequencies (above 4 kHz) and thus the

---

[25] $\mathbf{c}_{max}[0]$ is the maximum value of $\mathbf{c}[0]$ in the utterance.

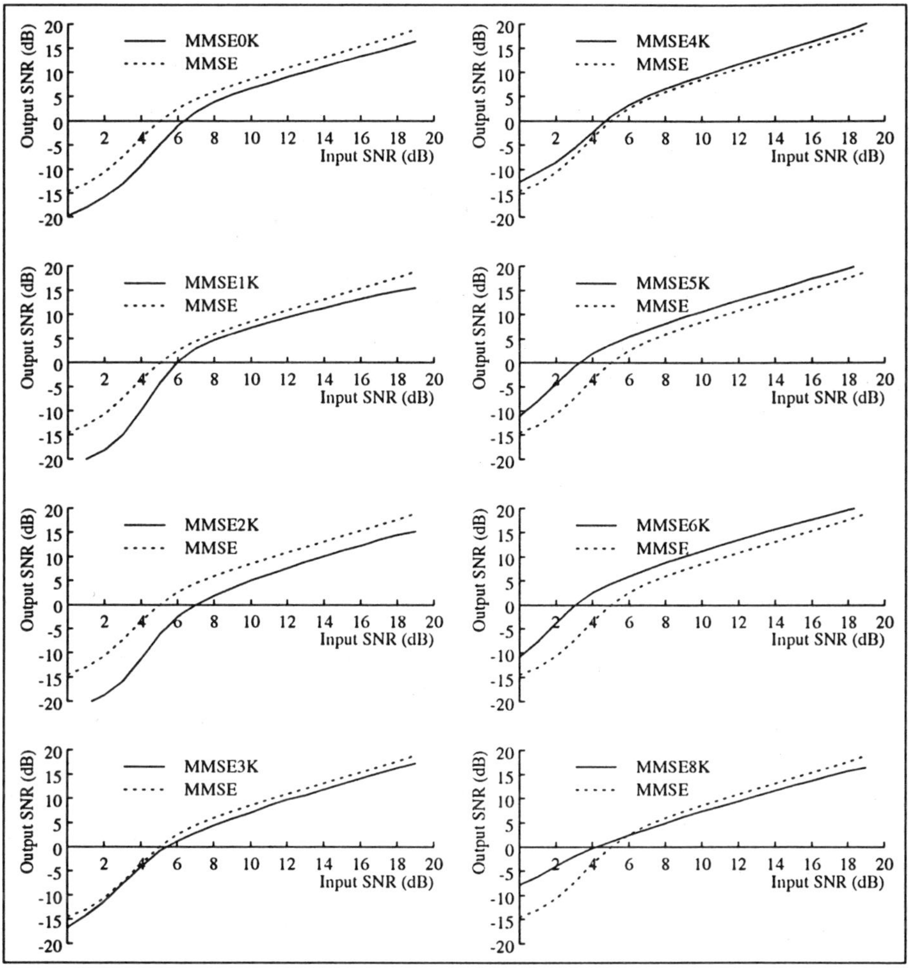

Figure 4-2: Comparison between the transformation curve MMSE fixed for all frequencies and the corresponding transformations for different frequencies: 0, 1, 2, 3, 4, 5, 6 and 8 kHz. The curves give the input-output relation between the SNR at a frequency band. It can be seen that more noise subtraction is done at low frequencies than at high frequencies. Also, low frequencies are attenuated more at high SNR, to compensate for spectral tilt.

algorithm is able to handle the colored noise present in the CRPZM speech. As far as spectral tilt is concerned, low frequencies (below 4 kHz) are attenuated while high frequencies (above 4 kHz) are boosted and therefore channel equalization is achieved.

The results of this new MMSEN algorithm are presented in Table 4-1. Since only the CRPZM speech is modified, the most important figure for comparison purposes is the one obtained with CLSTK training and CRPZM testing. We believe that the MMSEN performs better than the EQUAL + MMSE1 algorithm because it deals better with the colored noise and it provides spectral equalization as well as noise subtraction in a *joint* fashion (Compare Figures 3-10 and 4-3, especially at time 190 ms). Actually, application of matched pairs tests, McNemar's test and analysis of variance by ranks (Gillick [28], Pallett *et al.* [66]) showed that the difference between 61.3% and 66.4% is statistically significant with a confidence level higher than 95%.

TRAIN TEST	**CLSTK CLSTK**	**CLSTK CRPZM**	**CRPZM CLSTK**	**CRPZM CRPZM**
BASE	85.3%	18.6%	36.9%	76.5%
EQ+MMSE1	N/A	61.4%	75.8%	74.3%
MSUB	N/A	63.6%	71.7%	71.3%
MMSEN	N/A	66.4%	75.5%	72.3%

Table 4-1: Recognition accuracy of the MMSEN compared with the Baseline and the MMSE1 and MSUB algorithms.

The MMSEN algorithm performs also moderately better than the MSUB algorithm, although the difference is not statistically significant. Both algorithms use a family of curves as opposed to a single transformation curve, the difference is that in the MMSEN algorithm the dependence is on the frequency band while in the MSUB algorithm the dependence is on the input SNR. Conceivably a greater improvement could be attained if both dependencies are used together.

The same example that was presented in the previous chapter is shown in Figure 4-3 when processed by the MMSEN algorithm. The resulting noise is now approximately white, unlike the residual noise in the PSUB and MMSE1 algorithms, because of the different transformation used for every

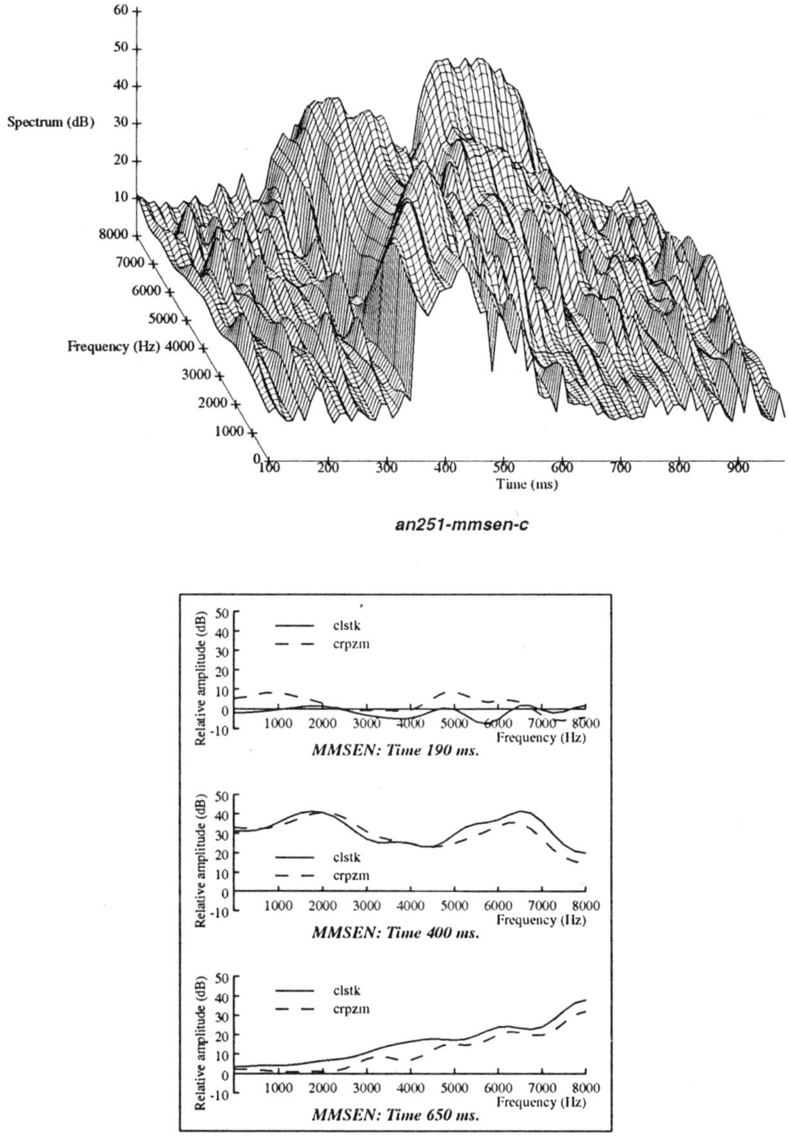

Figure 4-3: 3D Spectrogram of the utterance *yes* recorded with the CRPZM microphone with MMSEN algorithm. Spectra at times 190 ms (silence region), 400 ms (vowel) and 650 ms (fricative) are plotted for both the CLSTK and the CRPZM.

frequency band.

4.3. Processing in the Cepstral Domain: The SDCN Algorithm

Since SPHINX uses the LPC-cepstrum, implementation of the MMSEN algorithm requires two DFTs plus table look-ups (One DFT to go from the cepstral domain to the log-frequency domain, table look-ups for every frequency, and another DFT to return to the cepstral domain). The question that arises is whether it is possible to perform this normalization directly in the cepstral domain.

The *SNR-Dependent Cepstral Normalization* algorithm, SDCN, is an algorithm that operates directly in the cepstral domain by adding a compensation vector that depends exclusively on the SNR of the input frame. A justification for this idea within our model of the environment is given now.

Let **z** be a noisy estimate of **y** as defined in (4.8) obtained through our spectral estimation algorithm:

$$\mathbf{z} = \mathbf{y} + \mathbf{e} \tag{4.15}$$

where **e**, the estimation error, is a random vector. Our goal is to recover the uncorrupted vector **x** of an utterance given the observation **z** and our knowledge of the environment **n** and **q**.

$$\mathbf{x} = \mathbf{z} - \mathbf{e} - \mathbf{q} - \mathbf{r}(\mathbf{x}, \mathbf{n}, \mathbf{q}) \tag{4.16}$$

Unfortunately, given the non-linear expression (4.10), it is not possible to obtain a close-form solution for **x**. One possible approximation would be to assume that the compensation vector

$$\mathbf{w} = -\mathbf{e} - \mathbf{q} - \mathbf{r}(\mathbf{x}, \mathbf{n}, \mathbf{q}) \tag{4.17}$$

depends exclusively on the instantaneous SNR of the input frame:

$$\hat{\mathbf{x}} = \mathbf{z} + \mathbf{w}(SNR) \tag{4.18}$$

This is the basic equation for the SDCN algorithm.

Let's consider a physical interpretation of the algorithm. Although inspection of Equation (4.18) may lead to think that a crude approximation of Equation (4.10) has been made, we will show that the SDCN algorithm is asymptotically correct. At high SNR, inspection of Equations (4.3), (4.8) and (4.10) indicates that $\mathbf{x}[0] + \mathbf{q}[0] \gg \mathbf{n}[0]$, $\mathbf{r} \approx \mathbf{0}$, $\mathbf{y} \approx \mathbf{x} + \mathbf{q}$ and $\mathbf{w} \approx -\mathbf{q}$. On the other hand at low SNR, $\mathbf{x}[0] + \mathbf{q}[0] \ll \mathbf{n}[0]$, $\mathbf{y} \approx \mathbf{n}$ and $\mathbf{w} \approx \mathbf{m} - \mathbf{n}$ with $\mathbf{m}$ being the cepstrum of the noise for the clean signal. Hence, the SDCN algorithm performs spectral equalization at high SNR and noise suppression at low SNR. However, it is evident that at intermediate SNR, the SDCN algorithm can only be an approximation. Another simplification that we undertake is to estimate the input SNR as $\mathbf{z}[0] - \mathbf{n}[0]$. Although this is not the true instantaneous signal-to-noise ratio, it is related to it and it happens to be part of the feature vector of SPHINX.

We will now describe an MMSE method to estimate the compensation vectors $\mathbf{w}(SNR)$. The goal is to transform the CRPZM speech so that it looks like the CLSTK speech. Therefore according to this criterion, the correction vectors can be estimated by computing the average difference between cepstral vectors for the CRPZM speech versus the cepstral vectors for the CLSTK speech on a frame-by-frame basis as a function of the input SNR.

$$\mathbf{w}[j,k] = \frac{\sum_{i=0}^{N-1} (\mathbf{x}_i[j] - \mathbf{z}_i[j])\, \delta[SNR_i - k\Delta_{SNR}]}{\sum_{i=0}^{N-1} \delta[SNR_i - k\Delta_{SNR}]} \tag{4.19}$$

where $\mathbf{x}_i$ and $\mathbf{z}_i$ represent the cepstrum vectors at frame i for the CLSTK and CRPZM respectively. SNR_i is the SNR of frame i in the CRPZM. The sum is carried out for all the N frames in the database.

In general, the correction vectors can be estimated by computing the difference between the cepstral vectors of the test environment and the ones of a standard acoustical environment from simultaneous stereo recordings. The resulting correction vectors for our census database are shown in Figure 4-4.

Figure 4-5 shows the same example that was presented in the previous chapter when processed by the SDCN algorithm. We note that frames with low SNR (Compare frames at time 190 ms and 650 ms for Figures 4-3 and 4-5) are smoother in this case than in the MMSEN case.

The performance of the SDCN algorithm is shown in Table 4-2,

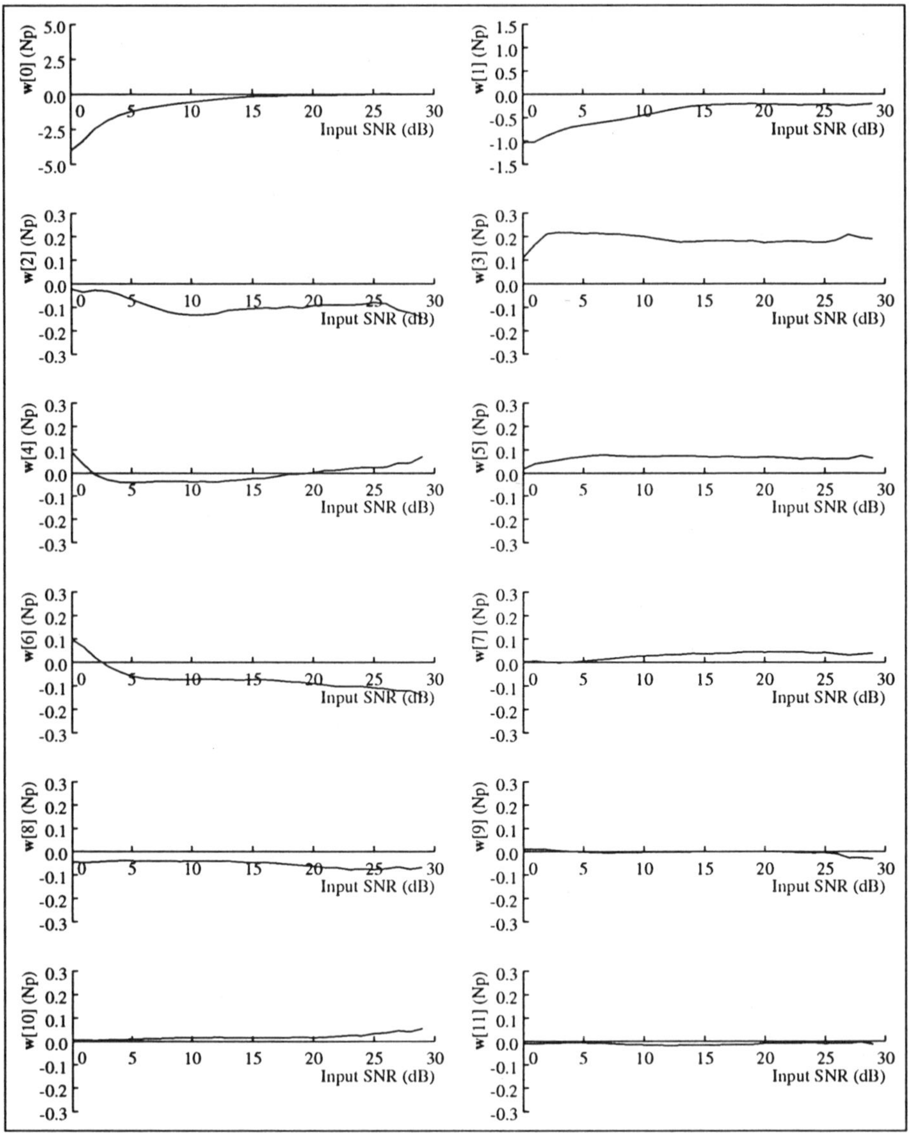

Figure 4-4: Correction vector $\mathbf{w}[i]$ in Np as a function of the instantaneous SNR in dB. Note the different scale used for $\mathbf{w}[0]$ and $\mathbf{w}[1]$, as they are the correction vectors varying the most. Correction vector $\mathbf{w}[12]$ is not shown.

together with the baseline condition and the previously described MMSEN.

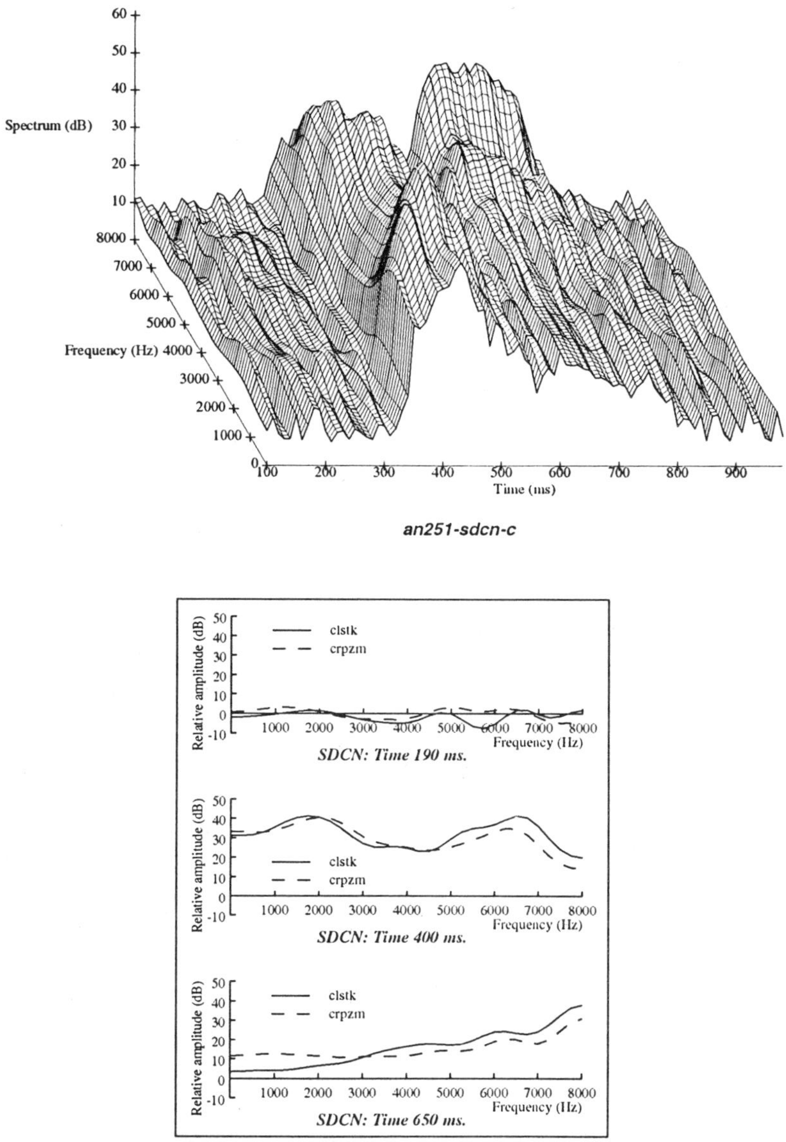

Figure 4-5: 3D Spectrogram of the utterance *yes* recorded with the CRPZM microphone with SDCN algorithm. Spectra at times 190 ms (silence region), 400 ms (vowel) and 650 ms (fricative) are plotted for both the CLSTK and the CRPZM.

This algorithm offers the highest accuracy of all the algorithms described so far. However, tests of statistical significance showed that the difference between 66.4% and 67.2% may be due to chance.

TRAIN TEST	CLSTK CLSTK	CLSTK CRPZM	CRPZM CLSTK	CRPZM CRPZM
BASE	85.3%	18.6%	36.9%	76.5%
MMSEN	N/A	66.4%	75.5%	72.3%
SDCN	N/A	67.2%	76.4%	75.5%

Table 4-2: Recognition accuracy of the MMSEN and SDCN algorithms when compared with the baseline.

As can be seen in Figure 4-4, the high-order components of **w** are small in magnitude and don't vary substantially with the SNR. Inspired by this fact, an experiment was carried out with the SDCN algorithm in which only the first p components of **w** were used for the compensation. The results are shown in Figure 4-6 for the case of training on the CLSTK and testing on the CRPZM. It is quite remarkable that only **w**[0] and **w**[1] are necessary. Use of the remaining components yields no further improvement in word accuracy.

4.4. Summary

In this chapter we have addressed the problem of *joint* compensation for noise and spectral tilt. The two algorithms presented, MMSEN and SDCN, successfully achieve the compensations in the frequency domain and the cepstral domain respectively. Both algorithms exhibit a higher accuracy than the algorithms that perform independent compensation for noise and spectral tilt described in Chapter 3.

There are some similarities between the two algorithms. They both make use of several transformation curves, one for each dimension in the space. The dependency on the MMSEN algorithm is on the channel SNR while for the SDCN algorithm it is on the frame SNR[26]. Typically,

[26]The instantaneous channel SNR is defined as the difference between the log-energy of the signal and the log-energy of the noise for a frequency band at a reference time. The frame SNR is defined as the difference between the **c**[0] of the input frame and the **c**[0] of the noise.

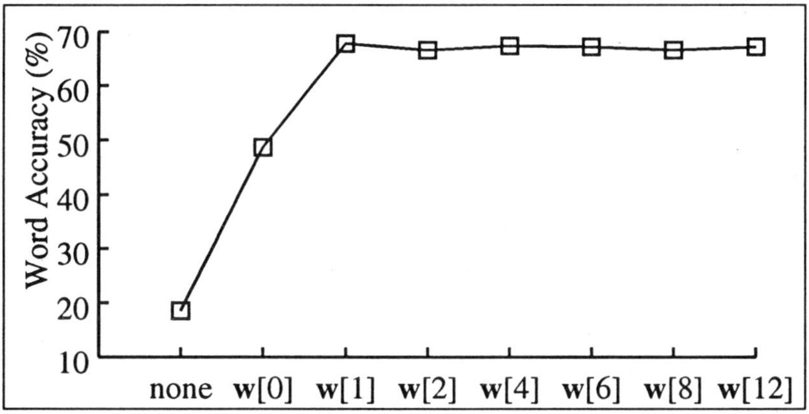

Figure 4-6: Word accuracy of the SDCN algorithm as a function of the number of correction vectors used when trained on the CLSTK and tested on the CRPZM. None means that no correction is applied (baseline), **w**[0] means that only **c**[0] was compensated, **w**[1] that both **c**[0] and **c**[1] were compensated, etc.

approaches that work in the frequency domain have a larger dimension (N = 32 in our case) than the those that operate in the cepstral domain (N = 12 in our case).

The curves for the MMSEN algorithm could be viewed as transformation curves as in Figure 4-2, or alternatively as an additive correction on the input value. Given the cepstral structure, SDCN can be viewed as strictly an additive correction (**w**[*i*] does not depend on **c**[*i*], just on the frame SNR).

Finally, both SDCN and MMSEN seem to perform equally well, but SDCN requires only 2 corrections whereas MMSEN needs 32. In addition, SDCN is more amenable for SPHINX since it operates on cepstrum vectors directly.

Even though the performance of the SDCN algorithm is quite acceptable, it requires a stereo database of our standard environment and the new environment to train the correction vectors. Since in a real situation such a database may not be available, the SDCN algorithm is not *environment-independent*.

We have also found that the above techniques produce many output frames that do not constitute legitimate speech vectors, especially at low SNR, because they do not take into account correlations across frequency bands or cepstral components. In the next chapter we discuss algorithms that explicitly consider the spectral profile of the compensated speech and that do not need a stereo database.

5 The CDCN Algorithm

Although the SDCN technique performs acceptably, it has the disadvantage that new microphones must be "calibrated" by collecting long-term statistics from a new stereo database. Since this stereo database will not be available in general, SDCN cannot adapt to a new environment. A new algorithm, *Codeword-Dependent Cepstral Normalization* (CDCN), was proposed to circumvent these problems, and will be the topic of this chapter.

In CDCN, the noise and channel equalization vectors are estimated so as to best match the acoustic space of the input speech frames with the acoustic space obtained by transforming a *universal* space with the environmental parameters, as depicted in Figure 5-1. This universal acoustic space is defined as the distribution of speech frames under a normalized clean environment, and is represented by a codebook of cepstrum vectors. A maximum likelihood criterion is used to estimate the environmental parameters that transform the standard acoustical ambiance into the acoustic space of the current environment.

Although the MMSEN and the SDCN algorithms were able to normalize *jointly* for the effects of noise and spectral tilt, the corrections are the same for all speech vectors. The MMSEN algorithm assumed that different frequency bands were uncorrelated, and that sometimes produced restored vectors that were not legitimate speech vectors, especially at low SNR. Likewise, the SDCN algorithm assumes that different cepstral components are uncorrelated, and it produced restored cepstral vectors that were not legitimate speech cepstrum vectors, again especially at low SNR.

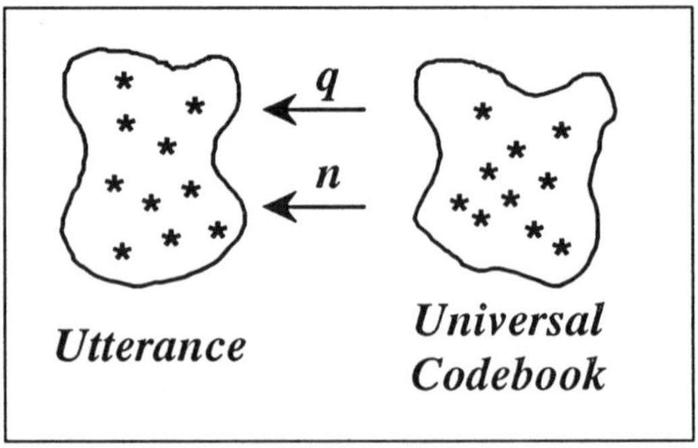

Figure 5-1: CDCN estimates the noise **n** and channel equalization **q** that best transform the universal codebook into the ensemble of input frames of the current utterance.

This use of *a priori* information for the speech vector was simultaneously proposed by Erell and Weintraub [19] and Acero and Stern [1]. Erell and Weintraub modeled the vector in the spectral domain as a mixture of Gaussians while Acero and Stern used a mixture of Gaussians in the cepstral domain. These models remove the assumption of independence across frequency bands (like the MMSEN) or across cepstral coefficients (like the SDCN).

These two approaches are similar, one working in the spectral domain and the other in the cepstral domain. Both of them clustered the acoustic space so that every cluster was transformed differently. In both cases an MMSE criterion was used for the restored vector that was essentially a weighted sum of the vectors transformed by all the different cluster transformations. One major difference is that in Erell's work only the noise suppression was performed with the above criterion and the spectral equalization was done separately, while in the CDCN algorithm both are done *jointly*. The other difference is that in Erell's work the estimation of the environment and especially the channel equalization, is performed by computing long-term averages, so it is not able to *adapt* to a new environment as the CDCN algorithm does. However, there is no fundamental reason why processing in the frequency domain could not perform joint compensation and adaptation to new environments.

First we define the framework we will be using and we outline the

CDCN algorithm with its two steps: MMSE estimation for the cepstrum and ML[27] estimation of the environmental parameters. Then we present a summary of the algorithm and evaluate the CDCN algorithm.

5.1. Introduction to the CDCN Algorithm

In this section we present the framework of the CDCN algorithm as well as the approximations that will be used. We start with a general MAP approach and then justify the decoupling of the estimation of the environmental parameters from the string decoding process. We also discuss the effect of using discrete HMMs as SPHINX does, rather than semi-continuous HMMs, especially at low SNRs.

The objective of the speech recognition system is to find the most likely string S and environmental parameters $\mathbf{n}$ and $\mathbf{q}$, from an ensemble of N cepstrum vectors $Z = \mathbf{z}_0,\ldots,\mathbf{z}_{N-1}$. We will assume that during the observation period, both $\mathbf{n}$ and $\mathbf{q}$ are constant, which is to say that the noise is assumed to be stationary during that segment of time. We also assume that the filter is not varying, so that S, $\mathbf{n}$ and $\mathbf{q}$ are chosen to maximize

$$p(S,\mathbf{n},\mathbf{q}\,|\,Z) = \frac{p(S,Z\,|\,\mathbf{n},\mathbf{q})\,p(\mathbf{n},\mathbf{q})}{p(Z)} \tag{5.1}$$

which is equivalent to maximizing $p(S,Z\,|\,\mathbf{n},\mathbf{q})$ if no *a priori* knowledge of $\mathbf{n}$ and $\mathbf{q}$ is available. By assuming independence across frames, $p(S,Z\,|\,\mathbf{n},\mathbf{q})$ can be expressed as

$$p(S,Z\,|\,\mathbf{n},\mathbf{q}) = p(S\,|\,\mathbf{n},\mathbf{q})\,p(Z\,|\,S,\mathbf{n},\mathbf{q}) = p(S)\prod_{i=0}^{N-1} p(\mathbf{z}_i\,|\,S,\mathbf{n},\mathbf{q}) \tag{5.2}$$

since $p(S\,|\,\mathbf{n},\mathbf{q})$ is not a function of $\mathbf{n}$ and $\mathbf{q}$.

One of the reasons the spectral subtraction schemes did not perform well is that some of the processed frames were not legitimate speech vectors, especially at low SNR. This occurred because no use was made of *a priori* information about the speech. A widely used assumption for the *pdf* of the cepstrum $\mathbf{x}$ is a mixture of K Gaussian densities with means $\mathbf{c}[k]$, covariance

[27]ML, Maximum Likelihood of the desired estimate, is a procedure used in estimation theory that is based on the maximization of the probability density function with respect to a parameter.

matrices Σ_k, and weights $P[k]$:

$$p(\mathbf{x}) = \sum_{k=0}^{K-1} P[k]\, p(\mathbf{x} \mid k) = \sum_{k=0}^{K-1} P[k]\, N_{\mathbf{x}}(\mathbf{c}[k], \Sigma_k) \qquad (5.3)$$

Under these model assumptions $p(\mathbf{z}_i \mid S, \mathbf{n}, \mathbf{q})$ is given by

$$\begin{aligned} p(\mathbf{z}_i \mid S, \mathbf{n}, \mathbf{q}) &= \sum_{k=0}^{K-1} \int p(\mathbf{z}_i, \mathbf{x}_i, k \mid S, \mathbf{n}, \mathbf{q})\, d\mathbf{x}_i \\ &= \sum_{k=0}^{K-1} \int p(\mathbf{z}_i \mid \mathbf{x}_i, k, S, \mathbf{n}, \mathbf{q})\, p(\mathbf{x}_i, k \mid S, \mathbf{n}, \mathbf{q})\, d\mathbf{x}_i \\ &= \sum_{k=0}^{K-1} \int p(\mathbf{z}_i \mid \mathbf{x}_i, k, \mathbf{n}, \mathbf{q})\, p(\mathbf{x}_i \mid k) P_i[k \mid S]\, d\mathbf{x}_i \end{aligned} \qquad (5.4)$$

since $p(\mathbf{z}_i \mid \mathbf{x}_i, k, S, \mathbf{n}, \mathbf{q})$ is not a function of the string S and $p(\mathbf{x}_i \mid k, S, \mathbf{n}, \mathbf{q})$ does not depend on $\mathbf{n}$ and $\mathbf{q}$ either.

For a vector $\mathbf{x}_i$ coming from mixture k given $\mathbf{n}$ and $\mathbf{q}$, we can obtain $\mathbf{y}_i$ as defined by Equation (4.8). In Appendix IV we model $\mathbf{z}_i$ as a Gaussian random vector $N_{\mathbf{z}_i}(\mathbf{y}_i, \Gamma)$, due to a finite data sample in our spectral estimation. In Appendix V.2 we show that the integral in (5.4) can be approximated by

$$\int p(\mathbf{z}_i \mid \mathbf{x}_i, k, \mathbf{n}, \mathbf{q})\, p(\mathbf{x}_i \mid k) d\mathbf{x}_i = N_{\mathbf{z}_i}(\mathbf{q} + \mathbf{r}[k] + \mathbf{c}[k], \Gamma + \Sigma_k) \qquad (5.5)$$

by assuming that the correction vector $\mathbf{r}[\mathbf{x}, k] = \mathbf{r}[k]$ is constant within vectors of the same mixture. This approximation is correct in the asymptotic sense, and for a finite number of mixture components was also found to be reasonable in practice. Using (5.5), Equation (5.2) can be expressed as

$$p(S, Z \mid \mathbf{n}, \mathbf{q}) = p(S) \prod_{i=0}^{N-1} \sum_{k=0}^{K-1} N_{\mathbf{z}_i}(\mathbf{q} + \mathbf{r}[k] + \mathbf{c}[k], \Gamma + \Sigma_k)\, P_i[k \mid S] \qquad (5.6)$$

Therefore, the problem of the recognition of a sequence of corrupted vectors can be solved by integrating the acoustic model $p_{\mathbf{z}_i}(\mathbf{z}_i \mid k, \mathbf{n}, \mathbf{q})) = N_{\mathbf{z}_i}(\mathbf{q} + \mathbf{r}[k] + \mathbf{c}[k], \Gamma + \Sigma_k)$, the HMM model $P_i[k \mid S]$, and the grammar information $P(S)$ in a semi-continuous HMM framework.

5.1.1. ML Estimation Based on Acoustic Information

Finding the ML estimates of $\mathbf{n}$ and $\mathbf{q}$ in Equation (5.6) requires a complicated search involving the HMM models and the grammar, that is, a joint optimization of the string S and $\mathbf{n}$ and $\mathbf{q}$. We aim to solve the maximization without resorting to the HMM models, just considering the acoustic information. Although this will not be as optimal, we opted for this solution for expediency. We define

$$P_i[k] = E_S\{ p(S)\, P_i[k \mid S] \} \tag{5.7}$$

as the average mixture probabilities. In the absence of other information, these average probabilities will be the same for all i.

Unfortunately, finding the ML estimate of $\mathbf{n}$ in Equation (5.6) seems difficult since $\mathbf{n}$ is only implicitly present in the correction vectors $\mathbf{r}$. To make the presence of $\mathbf{n}$ explicit we will consider two different *phones* when taking the expectation in Equation (5.7): the noise model ν and the speech model ξ (any other *phone*). With this assumption

$$\begin{aligned} p(\mathbf{z}_i \mid \mathbf{n}, \mathbf{q}) &= E_S[p(S, \mathbf{z}_i \mid \mathbf{n}, \mathbf{q})] \\ &= p(\nu)\, p(\mathbf{z}_i \mid \mathbf{n}, \mathbf{q}, \nu) + p(\xi)\, p(\mathbf{z}_i \mid \mathbf{n}, \mathbf{q}, \xi) \\ &= p(\nu)\, p_i(0 \mid \nu)\, N_{\mathbf{z}_i}(\mathbf{n} + \mathbf{s}[0], \Gamma) \\ &+ p(\xi) \sum_{k=1}^{K-1} p_i(k \mid \xi)\, N_{\mathbf{z}_i}(\mathbf{q} + \mathbf{r}[k] + \mathbf{c}[k], \Gamma + \Sigma_k) \end{aligned} \tag{5.8}$$

where we have assumed that the HMM model for the noise is so sharp that contains only one mixture component. For that case we have used the alternate expression in terms of $\mathbf{n}$ and $\mathbf{s}$. If the noise is stationary within an utterance, one mixture component is sufficient to represent it, so that this is totally general. We have also set Σ_0 to 0 by assuming that the variability observed in the noise samples comes from the spectral estimation process exclusively.

5.1.2. MMSE Estimation and SPHINX

However,since SPHINX uses discrete HMM models, replacing the summation in (5.6) by only one term will lead to errors, especially at low SNRs. For high SNR the correction vectors are approximately zero and the mixtures $N_{\mathbf{z}_i}(\mathbf{q} + \mathbf{r}[k] + \mathbf{c}[k], \Gamma + \Sigma_k)$ are well separated, so that substituting the sum with the maximum is a good approximation. Unfortunately, when the SNR is low, the correction vectors are significant and many codewords in the original space are transformed into essentially the same vector, so that by picking the maximum we are actually underestimating the sum. This is also the reason why the variance of the MAP estimate in this case is so large for low SNR.

To solve this problem we opted to choose the MMSE estimate, so that a vector $\mathbf{x}_i$ is obtained for every $\mathbf{z}_i$ as that which minimizes the squared error. Our experiments confirmed that doing this is considerably better than picking the maximum in (5.6).

5.2. MMSE Estimator of the Cepstral Vector

In this Section, we will derive an MMSE estimator for the restored cepstral vector $\hat{\mathbf{x}}$ by using the mixture model. In this case we will assume that the environmental parameters $\mathbf{n}$ and $\mathbf{q}$ are known.

In Appendix V.1 it is shown that the probability density function $p(\mathbf{x}|\mathbf{z},\mathbf{n},\mathbf{q})$ has the form:

$$p(\mathbf{x}|\mathbf{z},\mathbf{n},\mathbf{q}) = \frac{\sum_{k=0}^{K-1} P[k]\, p(\mathbf{z}|\mathbf{x},\mathbf{n},\mathbf{q},k)\, p(\mathbf{x}|k)}{\sum_{k=0}^{K-1} P[k] \int p(\mathbf{z}|\mathbf{x},\mathbf{n},\mathbf{q},k)\, p(\mathbf{x}|k)\, d\mathbf{x}} \tag{5.9}$$

This *a posteriori* probabilitygives us all the information needed for a classification problem. An MAP estimator would imply in this case maximizing a sum of Gaussians. An approximate MAP estimate would pick the mixture component that maximizes $p(\mathbf{z}|\mathbf{x},\mathbf{n},\mathbf{q},k)\, p(\mathbf{x}|k)$. This kind of MAP estimates have a large variance at low SNRs, because many different vectors in the clean space are transformed into essentially the same vector in the noisy space. As expected, the recognition accuracy of the system using an MMSE estimate was far better than for the MAP estimate. So the MMSE estimate was used for further development of the method. From (5.9) it is easy to obtain the MMSE estimate as:

$$\hat{\mathbf{x}}_{MMSE} = E[\mathbf{x} \mid \mathbf{z}, \mathbf{n}, \mathbf{q}] = \frac{\sum_{k=0}^{K-1} P[k] \int \mathbf{x}\, p(\mathbf{z} \mid \mathbf{x}, \mathbf{n}, \mathbf{q}, k)\, p(\mathbf{x} \mid k)\, d\mathbf{x}}{\sum_{k=0}^{K-1} P[k] \int p(\mathbf{z} \mid \mathbf{x}, \mathbf{n}, \mathbf{q}, k)\, p(\mathbf{x} \mid k)\, d\mathbf{x}} \tag{5.10}$$

where $p(\mathbf{x} \mid k)$ is the *pdf* of the kth mixture component and $p(\mathbf{z} \mid \mathbf{x}, \mathbf{n}, \mathbf{q}, k)$ is the *pdf* of the spectral estimator $p(\mathbf{z} \mid \mathbf{y})$, both of which are assumed to be Gaussian random vectors.

Under the assumption that $p(\mathbf{z} \mid \mathbf{x}, \mathbf{n}, \mathbf{q}, k)$ and $p(\mathbf{x} \mid k)$ are Gaussian densities, the product of the two can be expressed as follows (See Appendix V.2):

$$p(\mathbf{z} \mid \mathbf{x}, \mathbf{n}, \mathbf{q}, k)\, p(\mathbf{x} \mid k)$$

$$= N_{\mathbf{x}}(\mathbf{b}(\mathbf{x}, k), (\Gamma^{-1} + \Sigma_k^{-1})^{-1})\, N_{\mathbf{z}}(\mathbf{q} + \mathbf{r}(\mathbf{x}) + \mathbf{c}[k], \Gamma + \Sigma_k) \tag{5.11}$$

where the vector $\mathbf{b}(\mathbf{x}, k)$ is given by

$$\mathbf{b}(\mathbf{x}, k) = \Sigma_k (\Sigma_k + \Gamma)^{-1} (\mathbf{z} - \mathbf{q} - \mathbf{r}(\mathbf{x})) + \Gamma (\Sigma_k + \Gamma)^{-1} \mathbf{c}[k] \tag{5.12}$$

and the quantity $d(\mathbf{x}, k)$ is defined as

$$d(\mathbf{x}, k) = (\mathbf{z} - \mathbf{q} - \mathbf{r}(\mathbf{x}) - \mathbf{c}[k])^T (\Gamma + \Sigma_k)^{-1} (\mathbf{z} - \mathbf{q} - \mathbf{r}(\mathbf{x}) - \mathbf{c}[k]) \tag{5.13}$$

Unfortunately, it is not possible to obtain a closed-form expression for the integrals in (5.10) even with (5.11) because $\mathbf{r}$ depends on $\mathbf{x}$ as given by (4.10), and some approximations have to be made. We will assume that the correction vector $\mathbf{r}$ is constant for all vectors in a mixture. The finer the partition of the space is, the better the approximation. For the mixture k, the correction vector is assumed to be

$$\mathbf{r}[k] = \mathbf{r}(\mathbf{c}[k]) \tag{5.14}$$

where the mean of the mixture is taken to compute the correction vector from (4.10). Equations (5.12) and (5.13) are transformed according to (5.14):

$$\mathbf{b}[k] = \Sigma_k (\Sigma_k + \Gamma)^{-1} (\mathbf{z} - \mathbf{q} - \mathbf{r}[k]) + \Gamma (\Sigma_k + \Gamma)^{-1} \mathbf{c}[k] \tag{5.15}$$

$$d[k] = (\mathbf{z} - \mathbf{q} - \mathbf{r}[k] - \mathbf{c}[k])^T (\Gamma + \Sigma_k)^{-1} (\mathbf{z} - \mathbf{q} - \mathbf{r}[k] - \mathbf{c}[k]) \tag{5.16}$$

With these approximations, the estimate in (5.10) has the form

$$\hat{\mathbf{x}}_{MMSE} = \sum_{k=1}^{K-1} f[k]\, \mathbf{b}[k] \tag{5.17}$$

as shown in Appendix V.3, where

$$f[k] = \frac{P[k]\, N_{\mathbf{z}}(\mathbf{q} + \mathbf{r}[k] + \mathbf{c}[k], \Gamma + \Sigma_k)}{\sum_{l=0}^{K-1} P[l]\, N_{\mathbf{z}}(\mathbf{q} + \mathbf{r}[l] + \mathbf{c}[l], \Gamma + \Sigma_l)} \tag{5.18}$$

where $f[k]$ has the interpretation of the *a posteriori* probability of mixture k given the acoustic information $\mathbf{z}$ for the environmental parameters $\mathbf{n}$ and $\mathbf{q}$, and $\mathbf{b}[k]$ is the conditional mean for mixture component k.

Alternatively, (5.17) can be expressed as

$$\hat{\mathbf{x}}_{MMSE} = \mathbf{z} - \mathbf{w} \tag{5.19}$$

where the compensation vector $\mathbf{w}$ is given by

$$\mathbf{w} = \sum_{k=0}^{K-1} f[k]\,[\Sigma_k (\Sigma_k + \Gamma)^{-1} (\mathbf{q} + \mathbf{r}[k]) + \Gamma (\Sigma_k + \Gamma)^{-1} (\mathbf{z} - \mathbf{c}[k])] \tag{5.20}$$

so that Equation (5.19) resembles the SDCN algorithm. In this case, however, the compensation $\mathbf{w}$ is a weighted sum of *codeword-dependent* corrections.

5.3. ML Estimation of Noise and Spectral Tilt

In this section we develop the ML estimates for the noise $\mathbf{n}$ and equalization function $\mathbf{q}$ from an ensemble of N cepstrum vectors $Z = \mathbf{z}_0, \ldots, \mathbf{z}_{N-1}$. This maximization is done using only the acoustic information for simplicity in the computations. Since no closed-form solution is available, we will present an iterative algorithm.

If no *a priori* information is given about the noise and equalization vectors $\mathbf{n}$ and $\mathbf{q}$, the optimum estimation method is maximum likelihood. The parameters $\mathbf{n}$ and $\mathbf{q}$ are chosen to maximize $p(Z|\mathbf{n},\mathbf{q})$. By assuming that different frames are independent from each other, we can use the expression

$$\ln p(Z|\mathbf{n},\mathbf{q}) = \sum_{i=0}^{N-1} \ln p(\mathbf{z}_i|\mathbf{n},\mathbf{q}) \tag{5.21}$$

whose maximization leads to

$$\nabla_n \ln p(Z|\mathbf{n},\mathbf{q}) = \sum_{i=0}^{N-1} \frac{\nabla_n p(\mathbf{z}_i|\mathbf{n},\mathbf{q})}{p(\mathbf{z}_i|\mathbf{n},\mathbf{q})} = \mathbf{0} \tag{5.22}$$

where a similar expression can be derived for the gradient of $\mathbf{q}$, where

$p(\mathbf{z}_i | \mathbf{n}, \mathbf{q})$ can be expressed according to Equation (5.8) as follows:

$$p(\mathbf{z}_i | \mathbf{n}, \mathbf{q}) = \alpha \frac{P[0]}{|\Gamma|^{1/2}} \exp(-\frac{1}{2}(\mathbf{z}_i - \mathbf{n})^T \Gamma^{-1} (\mathbf{z}_i - \mathbf{n}))$$

$$+ \alpha \sum_{k=1}^{K-1} \frac{P[k]}{|C_k|^{1/2}} \exp(-\frac{1}{2}(\mathbf{z}_i - \mathbf{q} - \mathbf{r}[k] - \mathbf{c}[k])^T C_k^{-1} (\mathbf{z}_i - \mathbf{q} - \mathbf{r}[k] - \mathbf{c}[k])) \qquad (5.23)$$

where α is a constant factor and $C_k = \Gamma + \Sigma_k$. The first term is expressed as a function of the noise $\mathbf{n}$ explicitly to reflect the fact that the noise codeword ($k=0$) is largely insensitive to $\mathbf{q}$, and depends mostly on $\mathbf{n}$. Similarly, the other codewords are largely insensitive to $\mathbf{n}$ and depend mostly on $\mathbf{q}$.

Since (5.22) and (5.23) lead to a highly nonlinear equation, we use a variant of the EM algorithm described in Appendix VI. Instead of maximizing the log-likelihood directly

$$L(\mathbf{n}, \mathbf{q}) = \sum_{i=0}^{N-1} \ln f_{\mathbf{z}}(\mathbf{z}_i; \mathbf{n}, \mathbf{q}) \qquad (5.24)$$

we can maximize a function U instead

$$U(\mathbf{n}, \mathbf{q}, \mathbf{n}', \mathbf{q}') = \sum_{i=0}^{N-1} E\{ \ln f_{\mathbf{xz}}(\mathbf{x}_i, k, \mathbf{z}_i; \mathbf{n}, \mathbf{q}) | \mathbf{z}_i; \mathbf{n}', \mathbf{q}' \} \qquad (5.25)$$

that is the expected value of the logarithm of the *joint* density of the observed data Z, the unobserved data X and the mixture component k given an estimate of the environmental parameters $\mathbf{n}'$ and $\mathbf{q}'$. With these definitions, the function U in Equation (5.25) has the form (See Appendix VII):

$$U = \sum_{i=0}^{N-1} \frac{\sum_{k=0}^{K-1} P[k] \int p(\mathbf{z}_i | \mathbf{x}_i, k, \mathbf{n}', \mathbf{q}') p(\mathbf{x} | k) \ln p(\mathbf{z}_i, \mathbf{x}_i, k | \mathbf{n}, \mathbf{q}) \, d\mathbf{x}_i}{\sum_{k=0}^{K-1} P[k] \int p(\mathbf{z}_i | \mathbf{x}_i, k, \mathbf{n}', \mathbf{q}') p(\mathbf{x}_i | k) \, d\mathbf{x}_i} \qquad (5.26)$$

that is given as a function of the old parameter vectors $\mathbf{n}'$ and $\mathbf{q}'$. The maximization of (5.26) gives us new estimates for $\hat{\mathbf{n}}$ and $\hat{\mathbf{q}}$ that can be taken as initial values for the next iteration. The new estimates have the form (see Appendix VII)

$$\hat{\mathbf{n}} = \frac{\sum_{i=0}^{N-1} f_i[0]\, \mathbf{z}_i}{\sum_{i=0}^{N-1} f_i[0]} \tag{5.27}$$

$$\hat{\mathbf{q}} = \frac{\sum_{i=0}^{N-1} \sum_{k=1}^{K-1} f_i[k]\, (\mathbf{z}_i - \mathbf{c}[k] - \mathbf{r}'[k])}{\sum_{i=0}^{N-1} \sum_{k=1}^{K-1} f_i[k]} \tag{5.28}$$

with $f_i[k]$ given by (5.18).

It is important to note that in this case, due to the quadratic form of U with both $\mathbf{n}$ and $\mathbf{q}$, there is just one maximum in the function. That is, the local and global maxima in this case are the same, under the assumptions of the model. In practice, different starting points always converged to the same point for some cases analyzed in detail. This concludes the derivation of the CDCN algorithm. In the following section we will summarize the results.

5.4. Implementation Details

To simplify Equations (5.15) and (5.18) required in (5.17), we will now make the assumptions described in Appendix IV, in which the spectral (cepstral) distance $d = \|\mathbf{z} - \mathbf{y}\|$ between the true PCD (Power Cepstral Density) $\mathbf{y}$ and the observed cepstrum $\mathbf{z}$ is assumed to exhibit a Gaussian distribution

$$f_{\mathbf{z}}(d) = \frac{1}{\sqrt{2\pi}\,\gamma} \exp\left(-\frac{d^2}{2\gamma^2}\right) \tag{5.29}$$

In this case the cepstral distance takes the following value

$$d[k] = \|\mathbf{z} - \mathbf{q} - \mathbf{r}[k] - \mathbf{c}[k]\| \tag{5.30}$$

Furthermore, we will make the same Gaussian assumption for the cepstral distance for $p(\mathbf{x} \mid k)$, where all codewords share the same variance, except for the noise codeword. In addition, σ^2 will include the contribution of the codeword variance and the cepstral estimation variance. With this, Equation (5.18) can be expressed as

$$f[k] = \frac{\frac{P[k]}{\sigma} \exp\left(-\frac{d^2[k]}{2\sigma^2} \right)}{\frac{P[0]}{\gamma} \exp\left(-\frac{d^2[0]}{2\gamma^2} \right) + \sum_{k=1}^{K-1} \frac{P[k]}{\sigma} \exp\left(-\frac{d^2[k]}{2\sigma^2} \right)} \quad \text{for } k \neq 0 \tag{5.31}$$

with $d[k]$ given by (5.30) and

$$f[0] = \frac{\frac{P[0]}{\gamma} \exp\left(-\frac{d^2[0]}{2\gamma^2} \right)}{\frac{P[0]}{\gamma} \exp\left(-\frac{d^2[0]}{2\gamma^2} \right) + \sum_{k=1}^{K-1} \frac{P[k]}{\sigma} \exp\left(-\frac{d^2[k]}{2\sigma^2} \right)} \quad \text{for } k=0 \tag{5.32}$$

with $d[0]$ given by

$$d[0] = \| \mathbf{z} - \mathbf{n} - \mathbf{s}[0] \| = \| \mathbf{z} - \mathbf{n} \| \tag{5.33}$$

in which we set the correction vector for the noise codeword $\mathbf{s}[0]$ to 0.

We will assume that $\sigma^2 >> \gamma^2$, which was confirmed experimentally. With this, Equation (5.15) takes the form

$$\begin{aligned} \mathbf{b}[k] &= \mathbf{z} - \mathbf{q} - \mathbf{r}[k] \quad \text{for } k \neq 0 \\ \mathbf{b}[0] &= \mathbf{z} - \mathbf{n} \quad \text{for } k = 0 \end{aligned} \tag{5.34}$$

As pointed out in Appendix IV, γ^2 can be estimated by

$$\gamma^2 = \frac{1}{N} \sum_{i=0}^{N-1} \sum_{j=0}^{p} (\mathbf{z}_i[j] - \mathbf{c}[j])^2 \tag{5.35}$$

where $\mathbf{z}_i[j]$ represents the j coefficient of frame i of the observed cepstral vector. The selection of $\mathbf{c}$ can be done in two different ways depending on the interpretation of γ:

1. Cepstral estimation error. The vectors $\mathbf{z}_i$ are obtained by computing the cepstral coefficients for synthetically generated white Gaussian noise. In this case vector $\mathbf{c}$ is obtained as the LPC-cepstrum from the true AR coefficients of the synthetic noise. The resulting γ obtained by this method was 0.2 .

2. Noise variance. The vectors $\mathbf{z}_i$ are the noise vectors in the census database, and $\mathbf{c}$ represents the average noise vector. These noise vectors are selected as those whose zeroth component is below a threshold. The resulting γ obtained by this method was 0.3, which was the value selected here. This value reflects not only the cepstral estimation error, but also possibly a small non-stationarity in the noise, and model

inaccuracies.

Two constants were used to initialize the algorithm: the noise threshold and the average SNR. The noise threshold[28] was used to obtain an initial estimate of **n** (we set it to 5 dB) whereas the average SNR[29] was used to obtain an initial estimate of **q**[0] (we set it to 55 dB) with the other components of **q** set to 0. These values are not critical because CDCN will reestimate both **n** and **q**. Nevertheless, the choice of better initial estimates will speed up the convergence.

In addition, all $P[k]$ are considered identical, except for $P[0]$ which represents the fraction of noise frames in the database. Therefore, $P[0]$ was set to 0.25, while $P[k] = (1 - P[0]) / (K - 1)$ according to Equations (5.7) and (5.8). We have run CDCN on different databases and observed that the algorithm works properly even if the ratio of noise frames is different though.

The codebook elements $\{\mathbf{c}[k]\}$ are estimated from CLSTK speech with a standard Lloyd algorithm [52], in which the CDCN algorithm was embedded (*i.e.* every time **c**[k] were reestimated for a given K, CDCN was run on all input utterances). In this implementation the codebook contained 128 vectors[30]. The value of σ was obtained as the average distortion[31] in the iteration that reestimated the codebook vectors; the final value for 128 vectors was 0.54. Increasing the number of codewords to 256 yielded essentially no increase in word accuracy.

The initial estimate for the noise is the average of frames whose power is below a threshold (the same computation used in the algorithms of Chapter

---

[28] An initial estimate of the noise vector **n** is obtained by averaging all frames whose power, zeroth order cepstral coefficient, are between the minimum power in the utterance and the minimum power plus a threshold.

[29] The average SNR in an utterance is defined in this case to be the difference between the average power (zeroth order cepstral coefficient) for speech frames and the average power (zeroth order cepstral coefficient) for noise frames.

[30] Note that this codebook is different from all of SPHINX's codebooks because it includes both cepstrum and the power term.

[31] Equation (5.35) was used where $\mathbf{z}_i$ were all frames in the database and **c** was the closest vector in the codebook.

3). The equalization vector is initialized to zero, although if some *a priori* information is available it could be used as an initial estimate.

5.5. Summary of the CDCN Algorithm

We present here a summary of the CDCN algorithm with its two phases: estimation of the environmental parameters $\mathbf{n}$ and $\mathbf{q}$, and estimation of the compensated speech given $\mathbf{n}$ and $\mathbf{q}$.

ENVIRONMENT ADAPTATION achieved by ML estimation of the environmental parameters $\mathbf{n}$ and $\mathbf{q}$ is done as follows:

1. Assume initial values of $\hat{\mathbf{n}}^{(0)}$ and $\hat{\mathbf{q}}^{(0)}$ for $j = 0$.
2. **Estimate** the correction vectors $\hat{\mathbf{r}}^{(j)}[k]$ for $k=1,2,\ldots,K-1$; given $\hat{\mathbf{n}}^{(j)}$, $\hat{\mathbf{q}}^{(j)}$, and $\mathbf{x}=\mathbf{c}[k]$ according to (4.10):

$$\hat{\mathbf{r}}^{(j)}[k] = \mathrm{IDFT}\ \{\ln\ (1 + e^{\mathrm{DFT}\ [\hat{\mathbf{n}}^{(j)} - \hat{\mathbf{q}}^{(j)} - \mathbf{c}[k]]})\} \tag{5.36}$$

and the *a posteriori* probabilities for the mixtures $f_i[k]$ as

$$f_i[k] = \frac{\frac{P[k]}{\sigma}\exp(-\frac{d_i^2[k]}{2\sigma^2})}{\frac{P[0]}{\gamma}\exp(-\frac{d_i^2[0]}{2\gamma^2}) + \sum_{k=1}^{K-1}\frac{P[k]}{\sigma}\exp(-\frac{d_i^2[k]}{2\sigma^2})} \quad \text{for } k\neq 0 \tag{5.37}$$

$$f_i[0] = \frac{\frac{P[0]}{\gamma}\exp(-\frac{d_i^2[0]}{2\gamma^2})}{\frac{P[0]}{\gamma}\exp(-\frac{d_i^2[0]}{2\gamma^2}) + \sum_{k=1}^{K-1}\frac{P[k]}{\sigma}\exp(-\frac{d_i^2[k]}{2\sigma^2})} \quad \text{for } k=0 \tag{5.38}$$

where the distances $d_i[k]$ are given by

$$d_i[0] = \|\mathbf{z}_i - \hat{\mathbf{n}}^{(j)}\| \tag{5.39}$$

$$d_i[k] = \|\mathbf{z}_i - \hat{\mathbf{q}}^{(j)} - \mathbf{c}[k] - \hat{\mathbf{r}}^{(j)}[k]\| \tag{5.40}$$

3. **Maximize** the log-likelihood (5.21). The new estimates for $\hat{\mathbf{n}}^{(j+1)}$ and $\hat{\mathbf{q}}^{(j+1)}$ are

$$\hat{\mathbf{n}}^{(j+1)} = \frac{\sum_{i=0}^{N-1} f_i[0]\,\mathbf{z}_i}{\sum_{i=0}^{N-1} f_i[0]} \tag{5.41}$$

$$\hat{\mathbf{q}}^{(j+1)} = \frac{\sum_{i=0}^{N-1} \sum_{k=1}^{K-1} f_i[k]\,(\mathbf{z}_i - \mathbf{c}[k] - \hat{\mathbf{r}}^{(j)}[k])}{\sum_{i=0}^{N-1} \sum_{k=1}^{K-1} f_i[k]} \tag{5.42}$$

4. Stop if convergence has been reached obtaining $\hat{\mathbf{n}}$ and $\hat{\mathbf{q}}$; otherwise increment j and go to Step 2.

MMSE ESTIMATION of $\mathbf{x}_i$.Once the ML estimation of **n** and **q** has converged, for every frame in the utterance, the clean speech vectors are estimated as:

$$\hat{\mathbf{x}}_i = \sum_{k=1}^{K-1} f_i[k]\,(\mathbf{z}_i - \hat{\mathbf{q}} - \hat{\mathbf{r}}[k]) \quad i=0, 1, \ldots, N-1; \tag{5.43}$$

Let's attempt to interpret Equations (5.41), (5.42) and (5.43). Equation (5.41) estimates the noise vector **n** by a weighted sum of all the vectors $\{\mathbf{z}_i\}$ with the weights representing the *a posteriori* probability of each vector being noise. Equation (5.42) estimates **q** as a weighted sum of the deviations between the vector $\mathbf{z}_i$ and the corresponding "corrected" codeword $\mathbf{c}[k] + \hat{\mathbf{r}}[k]$. The weights $f_i[k]$ represent again the *a posteriori* probability for the mixture k at frame i, or how close the input vector i is to mixture k in a probabilistic sense given the acoustics. Finally equation (5.43) includes a correction that is a sum of all the correction vectors weighted by the appropriate *a posteriori* probabilities.

5.6. Evaluation Results

Table 5-1 describes the recognition accuracy of the original SPHINX system with no preprocessing, and with the SDCN and CDCN algorithms. Use of the CDCN algorithm brings the performance obtained when training on the CLSTK and testing on the CRPZM to the level observed when the system is trained and tested on the CRPZM. Moreover, use of CDCN improves performance obtained when training and testing on the CRPZM to a level greater than the baseline performance.

TRAIN TEST	CLSTK CLSTK	CLSTK CRPZM	CRPZM CLSTK	CRPZM CRPZM
BASE	85.3%	18.6%	36.9%	76.5%
SDCN	N/A	67.2%	76.4%	75.5%
CDCN	85.3%	74.9%	73.7%	77.9%

Table 5-1: Comparison of recognition accuracy of SPHINX with no processing, SDCN and CDCN algorithms. The system was trained and tested using all combinations of the CLSTK and CRPZM microphones.

To compare the CDCN algorithm with the previous algorithms we include the 3D spectrograms of the same utterance in Figures 5-2 and 5-3 for the CLSTK and CRPZM respectively. We note that in this case the algorithm is applied to both recordings.

We can see that the noise level has been reduced greatly not just for the CRPZM but for the CLSTK as well. The tilt in the spectrum has been almost completely eliminated as can be observed by looking at the frame at time 400 ms (the segment with highest SNR in the utterance).

To confirm the ability of the CDCN algorithm to adapt to new environmental conditions, a series of tests was performed with the 5 new stereo speech databases described in Chapter 2. The test data were all collected after development of the CDCN algorithm was completed. In all cases the system was trained using the Sennheiser HMD224. The "second" microphones (with which the system was *not* trained) were:

- The Crown PCC160 desk-top phase-coherent cardioid microphone (CRPCC160). (This is the new DARPA "standard" desk-top microphone.)
- An independent test set using the Crown PZM6fs.
- The Sennheiser 518 dynamic cardioid, hand-held microphone (SENN518).
- The Sennheiser ME80 electret supercardioid stand-mounted microphone (SENNME80).

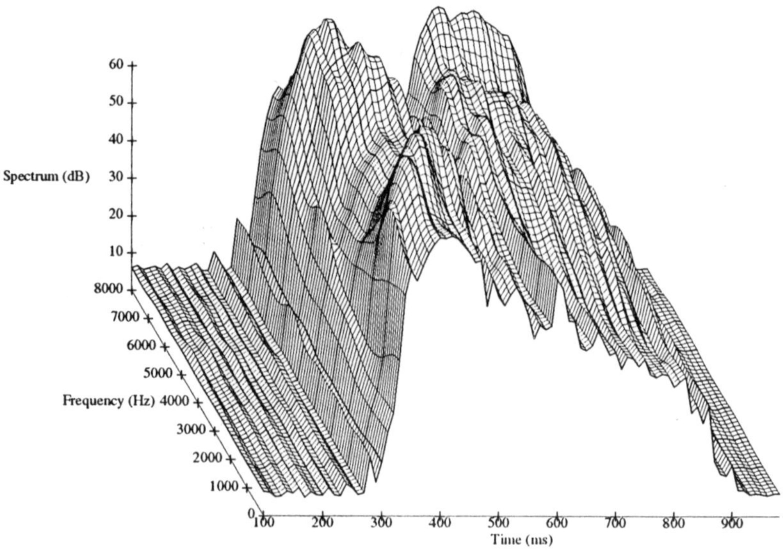

Figure 5-2: 3D Spectrogram of the utterance *yes* recorded with the CLSTK microphone processed with the CDCN algorithm.

- An HME lavalier microphone that also used an FM receiver (HME). In this case the other microphone used was not the CLSTK Sennheiser HMD224, but the Crown PCC160.

In Tables 5-1 and 5-2 we compare results using the CDCN algorithm to baseline performance. With this algorithm great robustness is obtained across microphones. However, there is a slight drop in performance when training and testing on the Sennheiser HMD224. We believe that one cause for this is that estimates of **q** and **n** are not very good for short utterances.

5.7. Summary

Although the SDCN algorithm performed acceptably, it had the drawback that a stereo database was required to train its correction vectors and hence it was not microphone-independent. The CDCN algorithm fits into an environment-independent framework because it estimates the parameters of the current environment via maximum likelihood. This capability of

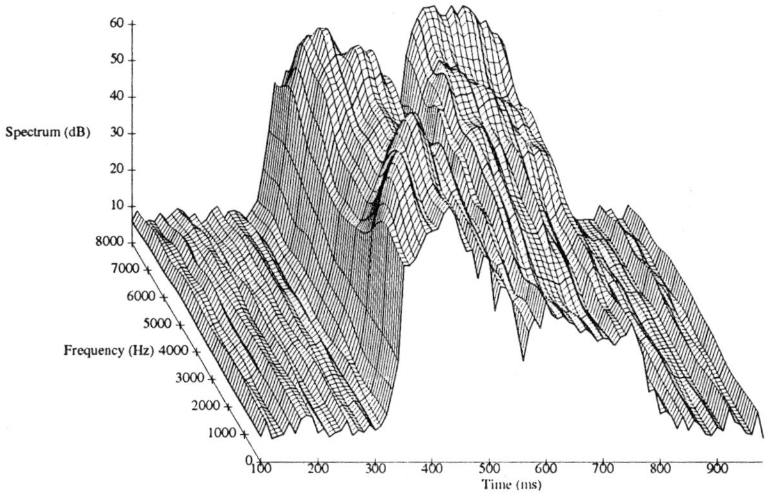

an251-cdcn-c

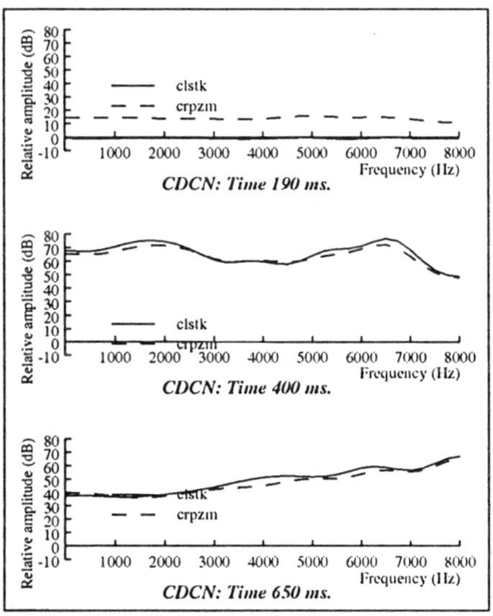

Figure 5-3: 3D Spectrogram of the utterance *yes* recorded with the CRPZM microphone with CDCN algorithm. Spectra at times 190 ms (silence region), 400 ms (vowel) and 650 ms (fricative) are plotted for both the CLSTK and the CRPZM.

TEST	CLSTK	CRPCC160
BASE	82.4%	70.2%
CDCN	81.0%	78.5%

TEST	CLSTK	CRPZM6FS
BASE	84.8%	41.8%
CDCN	83.3%	73.9%

TEST	CLSTK	SENN518
BASE	87.2%	84.5%
CDCN	82.2%	83.3%

TEST	CLSTK	SENNME80
BASE	83.7%	71.4%
CDCN	81.5%	80.7%

Table 5-2: Analysis of recognition accuracy of SPHINX for the baseline and the CDCN algorithm. Two microphones were recorded in stereo in each case. The microphones compared are the Sennheiser HMD224, 518, ME80, the Crown PZM6FS and the Crown PCC160. Training was done with the Sennheiser HMD224 in all cases.

TEST	HME	CRPCC160
BASE	55.9%	56.3%
CDCN	81.7%	72.2%

Table 5-3: Analysis of recognition accuracy of SPHINX for the baseline and the CDCN algorithm. Two microphones were recorded in stereo: the Crown PCC160, and the HME microphone. Training was done with the Sennheiser HMD224 in all cases.

adaptation to the environment makes the CDCN algorithm very attractive. The algorithm has been tested with a number of different microphones and in all cases showed great robustness.

The key information that allowed us to estimate the parameters of the environment, noise and spectral equalization, is the use of the *a priori* information about the speech cepstra contained in a codebook. By observing how that codebook is transformed under some noise **n** and equalization **q**, we were able to see the distribution of vectors under the current environment differed from the transformed distribution.

Another flaw of the SDCN algorithm was that the same correction was applied for all frames with the same SNR while it is clear that different frames would require different corrections. The CDCN algorithm uses corrections that are *codeword-dependent.*

The speech frames processed by the CDCN algorithm exhibit a much lower noise level, and the tilt in the spectrum is insignificant after the transformation. This shows that the model of the environment is quite accurate and that the CDCN algorithm is the first of the algorithms described in this monograph that can be considered to be *environment independent*, because the accuracy of the system when trained on speech recorded with a close-talking microphone and tested with speech recorded with a desk-top microphone is essentially equivalent to our benchmark (the accuracy of the system when trained *and* tested on speech recorded with the desk-top microphone).

6 Other Algorithms

In this chapter we present three additional algorithms: ISDCN, BSDCN and FCDCN. The ISDCN and BSDCN algorithms apply a compensation that is a function of the instantaneous SNR exclusively as in the case of SDCN, and they offer a similar recognition accuracy. ISDCN and BSDCN differ from the SDCN algorithm in the way the compensation vectors are estimated, which does not require a stereo database. The FCDCN algorithm is an extension of the SDCN algorithm that offers the highest recognition accuracy of all the algorithms proposed in this monograph, but as in the case of SDCN, it must be recalibrated for every new environment with a stereo database. Finally, we will study the real-time environmental adaptation capabilities of ISDCN, BSDCN and CDCN.

6.1. The ISDCN Algorithm

One of the deficiencies of the SDCN algorithm is the inability to adapt to new environments, since the correction vectors are derived from a stereo database of simultaneous recordings from our "standard" Sennheiser HMD224 and the new microphone. On the other hand, the algorithm is extremely simple and performs remarkably well. In this section we present the ISDCN algorithm [2], *Interpolated SNR-Dependent Cepstral Normalization*, as an extension of the SDCN algorithm to environment independence.

The correction vectors of the ISDCN algorithm will be a function of the instantaneous SNR like the SDCN, and the environmental parameters **n** and

q. Therefore, the compensated vector $\hat{\mathbf{x}}_i$ has the form

$$\hat{\mathbf{x}}_i = \mathbf{z}_i - \mathbf{w}(\mathbf{n}, \mathbf{q}, SNR_i) \tag{6.1}$$

This explicit dependence of **w** on **n** and **q** will allow us to estimate these parameters from the speech data by using the same criterion used in CDCN: minimization of the difference between the acoustic space of the current utterance and a universal acoustic space characterized by a codebook of cepstral vectors.

Although CDCN accomplishes this by maximum likelihood, we opted to use the minimization of the accumulated VQ distortion for ISDCN. Since SPHINX is already performing Vector Quantization, ISDCN can be implemented with very little computational overhead.

Given an ensemble of N frames, the accumulated VQ distortion has the form

$$D(\mathbf{n}, \mathbf{q}, k_i) = \sum_{i=0}^{N-1} \| \mathbf{z}_i - \mathbf{w}(\mathbf{n}, \mathbf{q}, SNR_i) - \mathbf{c}[k_i] \|^2 \tag{6.2}$$

where k_i is the label at frame i that minimizes the distortion

$$k_i = \text{argmin } \| \mathbf{z}_i - \mathbf{w}(\mathbf{n}, \mathbf{q}, SNR_i) - \mathbf{c}[k_i] \|^2 \tag{6.3}$$

In ISDCN the correction vector is expressed as

$$\mathbf{w}(\mathbf{n}, \mathbf{q}, SNR_i) = \mathbf{n} + (\mathbf{q} - \mathbf{n}) f(SNR_i) \tag{6.4}$$

where the function f *interpolates* between the noise **n** at low SNR and the equalization vector **q** at high SNR, so that the correction vector has the asymptotic characteristics noted in Chapter 4. We selected f to be the sigmoid function

$$f_i(x) = 1 / [1 + \exp(-\alpha_i x + \beta_i)] \quad \alpha_i > 0 \tag{6.5}$$

because it satisfies the asymptotic behavior of being $f \approx 0$ at low SNR and $f \approx 1$ at high SNR. It is also monotonic and very smooth.

The noise vector **n** can be reliably estimated by averaging a number of noise frames. We basically used the same procedure that we used in the spectral subtraction algorithms in Chapter 3: average all frames with a value of **c**[0] below a predetermined threshold.

Estimation of the equalization vector required the criterion used in Equation (6.2):

1. Start with an initial estimate for $\hat{\mathbf{q}}^{(0)}$ and $j = 1$
2. Label all frames, *i.e.* find the value of $k_i^{(j)}$ that minimizes the distortion

$$k_i^{(j)} = \operatorname{argmin} \|\mathbf{z}_i - \mathbf{w}(\mathbf{n}, \hat{\mathbf{q}}^{(j-1)}, SNR_i) - \mathbf{c}[k_i^{(j)}]\|^2 \quad i \le 0 \le N-1 \tag{6.6}$$

3. Estimate $\mathbf{q}^{(j)}$ from all the frames in the utterance:

$$\hat{\mathbf{q}}^{(j)} = \mathbf{n} + \frac{\sum_{i=0}^{N-1} (\mathbf{z}_i - \mathbf{n} - \mathbf{c}[k_i^{(j)}])\, f(SNR_i)}{\sum_{i=0}^{N-1} f^2(SNR_i)} \tag{6.7}$$

4. If convergence has been reached stop, else increment j and go to step 2.

It can be shown that the labels chosen in Step 2 that minimize Equation (6.6), will also minimize the overall distortion

$$D(\mathbf{n}, \hat{\mathbf{q}}^{(j-1)}, k_i^{(j)}) \le D(\mathbf{n}, \hat{\mathbf{q}}^{(j-1)}, k_i^{(j-1)}) \tag{6.8}$$

for a fixed $\mathbf{n}$ and $\hat{\mathbf{q}}^{(j-1)}$. Likewise, it can be shown that the new equalization vector selected by Equation (6.7) in Step 3 will reduce the distortion

$$D(\mathbf{n}, \hat{\mathbf{q}}^{(j)}, k_i^{(j)}) \le D(\mathbf{n}, \hat{\mathbf{q}}^{(j-1)}, k_i^{(j)}) \tag{6.9}$$

Therefore, this procedure is guaranteed to converge to a minimum of the accumulated distortion D, as every iteration will decrease the overall VQ distortion.

At this point we note that since we are using the codebook of cepstral vectors without the power term[32] (*i.e.* without $\mathbf{c}[0]$), the value of $\mathbf{q}[0]$ cannot be computed by Equations (6.6) and (6.7). The constraint we used to estimate $\mathbf{q}[0]$, the gain control, was that the dynamic range of every utterance ($\mathbf{c}_{max}[0] - \mathbf{c}_{min}$) had to be constant (We chose that constant to be 45 dB in our implementation).

---

[32]The codebook used here is SPHINX's cepstrum codebook that does not contain the power term.

For the evaluation, α_i and β_i were set empirically to 3.0 for $i > 0$ and 6.0 for $i=0$, as they were the values that made the sigmoids approximate most closely the curves in Figure 4-4. The equalization estimate $\hat{\mathbf{q}}$ given by Equation (6.7) exhibited a large variance for short utterances which introduced noise into the system. To ameliorate this problem, we only reestimate the first 4 cepstral coefficients of $\mathbf{q}$, setting to 0 the remaining coefficients. This reflects the fact that the equalization vector must be a smooth function. In Table 6-1 we show the performance of the census database with the ISDCN algorithm.

TRAIN TEST	CLSTK CLSTK	CLSTK CRPZM
BASE	85.3%	18.6%
CDCN	85.3%	74.9%
ISDCN	84.8%	62.1%

Table 6-1: Recognition accuracy of the ISDCN algorithm as compared with the baseline and the CDCN. The algorithm was applied to both CLSTK and CRPZM and training was done with the processed CLSTK.

Since ISDCN will estimate the environmental parameters $\mathbf{n}$ and $\mathbf{q}$ for every utterance, it does not have to be recalibrated for every new environment. An evaluation with the microphone recordings used in Chapter 5 is presented in Tables 6-5, 6-6, and 6-2. We see that the algorithm improves over the baseline in the cross conditions, although not as much as the CDCN for recordings with low SNR. For the case of CLSTK testing, the ISDCN algorithm does on the average about the same as the baseline, and slightly better than the CDCN algorithm.

6.2. The BSDCN Algorithm

In this section we present the Blind SDCN (BSDCN) algorithm as a variant of the SDCN algorithm described in Chapter 4. The BSDCN algorithm [54] processes corrupted speech just as SDCN does, but the compensation vectors are estimated in a way that does not require stereo recordings. We found the recognition accuracy of BSDCN to be very similar

TEST	HME	CRPCC160
BASE	55.9%	56.3%
CDCN	81.7%	72.2%
ISDCN	76.6%	69.7%

Table 6-2: Analysis of performance of SPHINX for the baseline and the CDCN and ISDCN algorithms. The Crown PCC160 and the HME FM microphone were recorded in stereo using training with the Sennheiser HMD224 (CLSTK) from the census database.

to that of SDCN with the advantage that BSDCN does not require stereo training.

In the BSDCN algorithm the need for stereophonic data is circumvented by lumping all data at each SNR together. A correspondence is established between SNRs in the training and testing environments by use of traditional nonlinear warping techniques [73] on histograms of SNRs from each of the two environments. The histograms of SNR values are first normalized for equal area, to avoid having the mapping be dominated by the environment from which more data had been collected. The minimum and maximum slopes of the warping path are limited to 0.2 dB/dB and 5 dB/dB, respectively, and the warping procedure seeks to minimize the Euclidean distance between the two histograms.

The SNR-warping procedure is illustrated in schematic form in Figure 6-1. The left and lower panels of Figure 6-1 show typical histograms of SNRs of speech collected using the PZM6FS and CLSTK microphones, respectively. The central panel of Figure 6-1 shows the warping path used to match SNRs from the two microphones. As can be seen in Figure 6-1, the mode in the SNR histogram for the CLSTK microphone at 26 dB is approximately matched to the mode in the SNR histogram for the PZM6FS microphone which actually occurs at 9 dB.

Since the alignment obtained by dynamically warpingthe histograms of the training and the testing data is not perfect, we have found that it is beneficial to smooth the correction vectors using the simple function

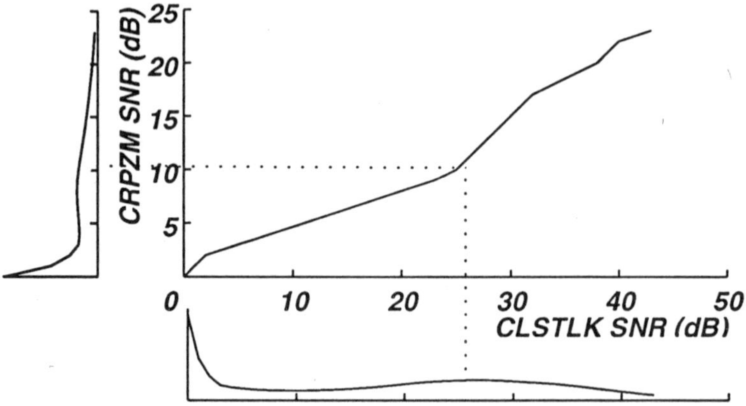

Figure 6-1: Nonlinear mapping of SNRs for the CLSTK and PZM6FS microphones based on histograms of SNR values. The unlabeled graphs along the horizontal and vertical axes indicate the relative likelihood of observing various instantaneous SNRs for the two microphones. The central panel indicates the warping path that best matches the two functions.

$$\hat{\mathbf{v}}(SNR) = 0.40\,\mathbf{v}(SNR) + 0.24\,\mathbf{v}(SNR+1)$$
$$+\ 0.24\,\mathbf{v}(SNR-1) + 0.06\,\mathbf{v}(SNR+2) + 0.06\,\mathbf{v}(SNR-2) \tag{6.10}$$

where **v** refers to an arbitrary cepstral vector from either environment, and SNR is in dB. Applying the above smoothing function to cepstra from both the training and testing data, we have reduced recognition error rate by an average of about 10 percent.

Once a correspondence is established between the SNRs in the training and testing environments, correction vectors are computed as the difference between average cepstra for every SNR in the testing environment and its corresponding SNR in the training environment.

Table 6-3 compares the recognition accuracy obtained when the BSDCN algorithm is evaluated using the alphanumeric census database. We note that the environment-independent BSDCN algorithm achieves a level of recognition accuracy when trained on the CLSTK microphone and tested on the PZM6FS microphone that is approximately equal to the recognition

TEST	CLSTK	PZM6FS
BASE	86.9%	31.4%
SDCN	N/A	72.4%
BSDCN	86.4	70.0%

Table 6-3: Recognition accuracy of the BSDCN algorithm compared with the baseline and SDCN, using testing data from the CLSTK and PZM6FS microphones. The system was trained using speech from the CLSTK microphone. This version of SPHINX had 4 codebooks that included the second order differences of cepstral coefficients.

accuracy achieved by the environment-dependent SDCN algorithm on the same task.

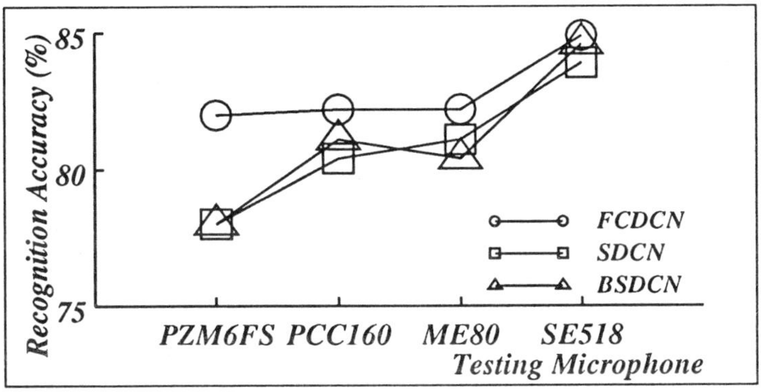

Figure 6-2: Comparison of recognition accuracy for SPHINX on the alphanumeric task using four microphones and the SDCN, BSDCN and FCDCN algorithms. The system was trained using speech from the CLSTK microphone. This version of SPHINX had 4 codebooks that included the second order differences of cepstral coefficients. For this new version, testing on the CLSTK yielded 86.9% and testing on the CRPZM 31.4%.

Figure 6-2 compares the recognition accuracy obtained using the BSDCN, SDCN, and FCDCN algorithms for four microphones: the omnidirectional desktop PZM6FS, the Crown PCC160 cardioid desktop microphone (PCC160), the Sennheiser ME80 supercardioid electret microphone (ME80), and the Sennheiser 518 handheld dynamic cardioid microphone (SE518). The system was trained with speech from the CLSTK microphone in all cases. We found again that accuracy obtained using the environment-independent BSDCN algorithm was comparable to that of the environment-dependent SDCN algorithm.

6.3. The FCDCN Algorithm

In this section we describe the Fixed CDCN algorithm (FCDCN) that combines some of the attractive features from both the SDCN and CDCN algorithms. The motivation for FCDCN is to obtain an algorithm that is as accurate as CDCN and as computationally efficient as SDCN.

The FCDCN algorithm [2] applies a correction that depends on the instantaneous SNR of the input, like SDCN, and this correction is different for every codeword, like CDCN:

$$\hat{\mathbf{x}} = \mathbf{z} + \mathbf{r}[k, SNR] \tag{6.11}$$

The selection of the appropriate codeword is done at the VQ stage ,so that label k is chosen to minimize

$$k = \text{argmin } \|\mathbf{z} + \mathbf{r}[k, SNR] - \mathbf{c}[k]\|^2 \tag{6.12}$$

This technique is applied only to the CRPZM speech so as to make it as close as possible to the CLSTK speech. We will describe a method for training these correction vectors using the EM algorithm (See Appendix VI). The incomplete data consist of the noisy speech, and the complete data include the clean speech and the mixture component, as it is known in the stereo database.

We can see that when K, the number of mixture components, equals 1, FCDCN reverts to the SDCN algorithm described in Chapter 4. In Figure 6-3 we show recognition accuracy as a function of K. It can be seen that an improvement can be obtained by using $K > 1$, though the amount of improvement saturates around $K = 4$ or 8.

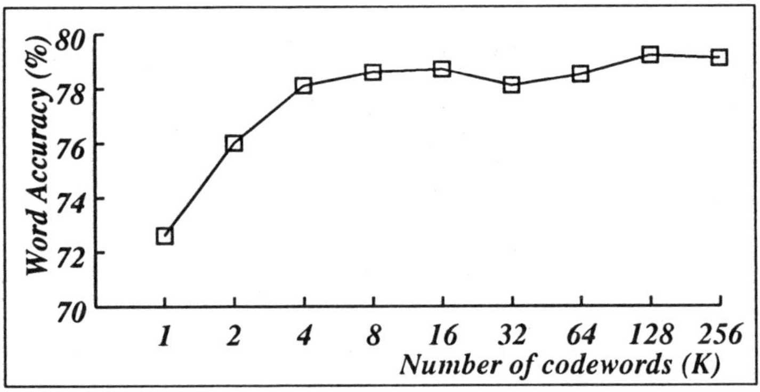

Figure 6-3: Word accuracy of the FCDCN algorithm as a function of the number of codewords K. Note that for $K = 1$, FCDCN reverts to the SDCN algorithm. Training is done on the CLSTK and testing on the CRPZM with the FCDCN algorithm. This version of SPHINX had 4 codebooks that included the second order differences of cepstral coefficients. For this new version, testing on the CLSTK yielded 86.9% and testing on the CRPZM 31.4%.

6.3.1. ML Estimation of the Correction Vectors

In this subsection we develop the estimation formulae for the correction vectors $\mathbf{r}[k, SNR]$ and the variances $\sigma^2[SNR]$ via the EM algorithm. Briefly, the correction vectors are reestimated as the vectors that minimize the differences between the CLSTK and the CRPZM cepstra on a frame-by-frame basis.

The densities to be used in the EM algorithm are defined here. As we did in Chapter 5, we will assume that the spectral distance of vector $\mathbf{x}$ to the mixture mean is a Gaussian variable. With this, the *pdf* of the CLSTK speech given mixture k is

$$p(\mathbf{x} \mid k) = \frac{1}{\sqrt{2\pi}\,\sigma} \exp(-\frac{1}{2\sigma^2} \| \mathbf{x} - \mathbf{c}[k] \|^2) \tag{6.13}$$

Now we will assume that the observed cepstra $\mathbf{z}$ is related to the clean cepstra $\mathbf{x}$ from mixture k by

$$\mathbf{z} = \mathbf{x} - \mathbf{r}[k, SNR] + e(SNR) \tag{6.14}$$

where the error $e(SNR)$ will be greater at low SNR, where there is more uncertainty in the knowledge about $\mathbf{x}$ that has been masked by the noise. We will further assume that this error is normally distributed. The CRPZM speech is therefore modeled as

$$p(\mathbf{z} \mid \mathbf{x}, k, \mathbf{r}', SNR = l) = \frac{1}{\sqrt{2\pi}\,\gamma[l]} \exp\left(-\frac{1}{2\gamma^2[l]} \|\mathbf{z} + \mathbf{r}'[k, l] - \mathbf{x}\|^2\right) \tag{6.15}$$

Multiplying (6.13) and (6.15), integrating with respect to $\mathbf{x}$ and making use of the Gaussian decomposition of Appendix V we obtain

$$p(\mathbf{z} \mid k, \mathbf{r}', SNR = l) = \frac{1}{\sqrt{2\pi}\,\sigma[l]} \exp\left(-\frac{1}{2\sigma^2[l]} \|\mathbf{z} + \mathbf{r}'[k, l] - \mathbf{c}[k]\|^2\right) \tag{6.16}$$

with $\sigma^2[l] = \sigma^2 + \gamma^2[l]$. In (6.15) and (6.16), $\mathbf{r}'[k, l]$ is the correction vector when the true mixture is k and the input SNR of $\mathbf{z}$ is $l\Delta_{SNR}$. The variance $\sigma[l]$ is a function of the instantaneous SNR too. With these assumptions, the*a posteriori probability* $p(k \mid \mathbf{z}_i, \mathbf{r}', SNR = l)$, that we will denote by $f_i[k]$ can be expressed as

$$f_i[k] = \frac{\exp\left(-\frac{1}{2\sigma^2[l_k]} \|\mathbf{z}_i + \mathbf{r}'[k, l_k] - \mathbf{c}[k]\|^2\right)}{\sum_{p=0}^{K-1} \exp\left(-\frac{1}{2\sigma^2[l_p]} \|\mathbf{z}_i + \mathbf{r}'[p, l_p] - \mathbf{c}[p]\|^2\right)} \tag{6.17}$$

where we have assumed that the *a priori* probabilities for the mixtures are identical. We can also obtain the logarithm of the complete data as

$$\ln p(\mathbf{z}_i, \mathbf{x}_i, k \mid \mathbf{r}) = C'' - \ln \gamma[l_k] - \frac{1}{2\gamma^2[l_k]} \|\mathbf{z}_i + \mathbf{r}[k, l_k] - \mathbf{x}_i\|^2 - \ln \sigma - \frac{1}{2\sigma^2} \|\mathbf{x}_i - \mathbf{c}[k]\|^2 \tag{6.18}$$

Instead of maximizing the likelihood $p(\mathbf{z}_i, \mathbf{x}_i \mid \mathbf{r})$ directly, we will use the EM algorithm described in Appendix VI and maximize the function U instead. In this case the function U has the form

$$U(\mathbf{r}, \mathbf{r}') = \sum_{i=0}^{N-1} \sum_{k=0}^{K-1} \ln p(\mathbf{z}_i, \mathbf{x}_i, k \mid \mathbf{r})\, p(k \mid \mathbf{z}_i, \mathbf{r}') \tag{6.19}$$

where the incomplete data are k and $\mathbf{x}_i$. Since $\{\mathbf{x}_i\}$ are known from stereo recordings, we used it directly. The correction vectors $\mathbf{r}[k, l_k]$ that maximize Equation (6.19) can be obtained by using Equations (6.17) and (6.18):

$$\mathbf{r}[k,l] = \frac{\sum_{i=0}^{N-1} (\mathbf{x}_i - \mathbf{z}_i)\, f_i[k]\, \delta[SNR_i - l\Delta_{SNR}]}{\sum_{i=0}^{N-1} f_i[k]\, \delta[SNR_i - l\Delta_{SNR}]} \tag{6.20}$$

and the corresponding $\gamma[l]$ given by

$$\gamma^2[l] = \frac{\sum_{i=0}^{N-1}\sum_{k=0}^{K-1} \| \mathbf{x}_i - \mathbf{z}_i - \mathbf{r}[k,l] \|^2 f_i[k]\, \delta[SNR_i - l\Delta_{SNR}]}{\sum_{i=0}^{N-1}\sum_{k=0}^{K-1} f_i[k]\, \delta[SNR_i - l\Delta_{SNR}]} \tag{6.21}$$

To simplify even more, we assumed that $\gamma[l] >> \sigma$ which was confirmed in practice. In fact, the recognition accuracy was not observed to be sensitive to the exact value of $\sigma[l]$. Figure 6-4 shows the resulting variances $\sigma^2[l]$ obtained after the process for $\Delta_{SNR} = 1\,dB$. The large variance exhibited at low SNR reflects the higher uncertainty in the value of the restored CLSTK speech given CRPZM speech for low SNRs.

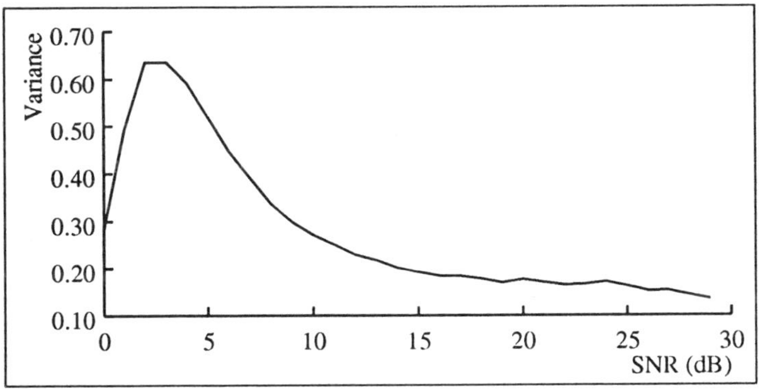

Figure 6-4: Variance of the difference vector between the CLSTK and the restored CRPZM speech for different input SNR of the CRPZM.

The new estimates $\mathbf{r}[k,l]$ and $\gamma[l]$ obtained by Equations (6.20) and (6.21) will increase the likelihood $p(Z,X|\mathbf{r})$. In practice, the algorithm reaches convergence after 2 or 3 iterations.

We have shown in Chapter 5 that what determines the values of the correction vectors is not just **q** but **q** – **n**, the difference between the equalization and noise vectors. Therefore in this implementation the first step consisted in subtracting the noise vector from all the input frames and then applying the correction vectors. In training the correction vectors are obtained the same way, subtracting the noise vector first. The noise vector is estimated via an EM procedure as described by Van Compernolle [11], in which two Gaussian densities are fitted to the data, one for the noise event and another one for the speech event. Table 6-4 shows the results of the FCDCN algorithm for the original set of recordings. In Tables 6-5 and 6-6 we show cross validation results with different microphone recordings. In Figure 6-5 we show the sample utterance for the CRPZM when processed by this algorithm.

TRAIN **TEST**	**CLSTK** **CLSTK**	**CLSTK** **CRPZM**
BASE	85.3%	18.6%
CDCN	85.3%	74.9%
FCDCN	N/A	73.1%

Table 6-4: Performance of the FCDCN algorithm as compared with the baseline and the CDCN. This algorithm is only applied to the CRPZM using the training models for the baseline case.

Another possibility for estimating a similar set of correction vectors was suggested by Gish *et al.* [29]. He used a procedure similar to Equation (6.20) for estimating the corrections but they derived the labels from the CLSTK speech instead of the CRPZM speech. We tried this procedure and obtained a restored speech that exhibited a large amount of *musical noise* as was observed using the PSUB algorithm (See Chapter 3). Many frames in the clean acoustic space are mapped into essentially the same frame in the noisy space because they are masked by the noise, so inverting the process causes many frames that are similar in the noisy acoustic space to become wildly different in the transformed space if this approach is taken. In our criterion the distortion measure is obtained in the clean space, so that small differences in the noisy speech gets translated into small differences in the restored speech.

TEST	CLSTK	CRPZM6FS
BASE	84.8%	41.8%
CDCN	83.3%	73.9%
ISDCN	86.1%	73.7%
FCDCN	N/A	79.3%

TEST	CLSTK	CRPCC160
BASE	82.4%	70.2%
CDCN	81.0%	78.5%
ISDCN	82.3%	75.4%
FCDCN	N/A	77.1%

TEST	CLSTK	SENN518
BASE	87.2%	84.5%
CDCN	82.2%	83.3%
ISDCN	87.2%	83.5%
FCDCN	N/A	83.4%

Table 6-5: Recognition accuracy of SPHINX for the baseline and the CDCN, ISDCN and FCDCN algorithms. Two microphones were recorded in stereo in each case. The microphones compared are the Sennheiser HMD224 (CLSTK), Crown PZM6FS, Crown PCC160 and Sennheiser 518. Training was done with the CLSTK from the census database in all cases. The correction vectors for FCDCN were estimated from each stereo database and are different for every experiment.

Another variant that we tried for estimating the correction vectors is using the assumption that $\mathbf{x}$ has the distribution

$$p(\mathbf{x}) = \max\ P[k]\ N_{\mathbf{x}}(\mathbf{c}[k], \Sigma_k) \tag{6.22}$$

where we have substituted the summation operator by the maximum operator. The *a posteriori* probability $p(k \mid \mathbf{z}_i, \mathbf{r}')$ will be a delta function $\delta[k]$

TEST	CLSTK	SENNME80
BASE	83.7%	71.4%
CDCN	81.5%	80.7%
ISDCN	83.2%	78.5%
FCDCN	N/A	81.1%

Table 6-6: Recognition accuracy of SPHINX for the baseline and the CDCN, ISDCN and FCDCN algorithms. Two microphones were recorded in stereo: the Sennheiser HMD224 (CLSTK) and SennheiserME80. Training was done with the CLSTK from the census database. The correction vectors for FCDCN were estimated from the stereo database.

where k is the codeword that minimizes

$$k = \text{argmin } \|\mathbf{z}_i - \mathbf{c}[k] - \mathbf{r}[k,l]\|^2 \tag{6.23}$$

Using the complete probability of Equation (6.18), our EM algorithm yields the same reestimation formula for $\mathbf{r}[k,l]$ given in Equation (6.20), with $f_i[k]$ being 1 if k is the label for frame i and 0 otherwise. The recognition rate for the CRPZM training on the CLSTK is 72.6%, slightly lower than with the previous training algorithm.

The computational complexity of this FCDCN is very low because the correction vectors are precomputed. However, it does not have the adaptation capabilities of the ISDCN or CDCN. It would be desirable to obtain an algorithm that is computationally efficient and adapts to the environment like the ISDCN but is more immune to noise. Having different sets of correction vectors for different environments (*i.e.* quantizing the environments) would be a possibility, but it would require a large number of different microphone recordings. Another possibility could be to interpolate between different environments.

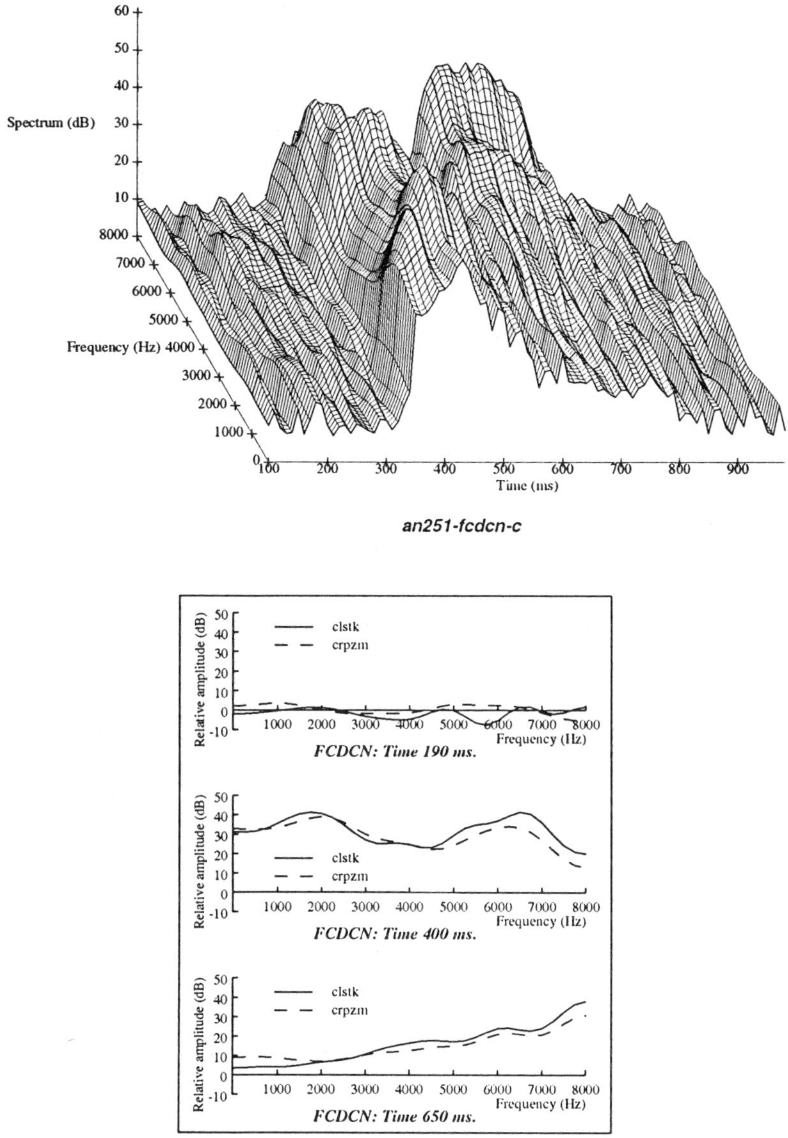

Figure 6-5: 3D Spectrogram of the utterance *yes* recorded with the CRPZM microphone with FCDCN algorithm. Spectra at times 190 ms (silence region), 400 ms (vowel) and 650 ms (fricative) are plotted for both the CLSTK and the CRPZM.

6.4. Environmental Adaptation in Real Time

In this section we analyze some issues related to real-time implementation. First we will describe an implementation of ISDCN that estimates the environmental parameters from the previous utterance with no loss of recognition accuracy. Then we present experiments of both CDCN and BSDCN, whose correction vectors were estimated from past samples of speech, and study the amount of speech needed for environmental adaptation.

Up to this point, the estimates of the noise **n** and equalization vector **q** for a given utterance were estimated from that same utterance. In a real application this would not be practical. Assuming that these parameters do not change very rapidly with time (quasi-stationarity), one should be able to use the estimates from the previous utterance for the next utterance.

Another test set was collected in stereo with the Sennheiser HMD224 and the CRPZM 6FS. In Table 6-7 we show the baseline performance for this set as well as the FCDCN and ISDCN algorithms. We see that this test set is probably the most difficult we have encountered since the performance for the case of training and testing on the CRPZM is only 66.6%. The accuracy of the FCDCN algorithm with the correction vectors computed from the census database exceeds that value, though the difference is not statistically significant.

TRAIN **TEST**	**CLSTK** **CLSTK**	**CLSTK** **CRPZM**	**CRPZM** **CLSTK**	**CRPZM** **CRPZM**
BASE	82.4%	21.3%	36.6%	66.6%
FCDCN	82.4%	67.2%	-	-
ISDCN	80.8%	57.1%	-	-
ISDCNprev	80.7%	55.8%	-	-

Table 6-7: Recognition accuracy of the ISDCN algorithm when the environmental parameters are computed from the same utterance (ISDCN) or the previous utterance (ISDCNprev). The performance is compared to the baseline (BASE) and the FCDCN algorithm.

Inspection of Table 6-7 tells us that use of the estimates of the environment computed from the previous utterance (ISDCNprev) yields essentially the same recognition accuracy than if the environmental parameters were estimated from the current utterance (ISDCN) as we have done until now. This fact can be used in implementing a real-time system where the estimates of **n** and **q** are continuously updated.

We have also produced a real-time implementation of the original CDCN algorithm. Although the CDCN algorithm is intrinsically more computationally costly than either the SDCN or FCDCN algorithms, we integrated a version of this algorithm into a real-time spoken language system [86] without any apparent additional processing time to the user. This was accomplished in two ways. First, the compensation and normalization parameters **n** and **q** are computed in the background during the silent intervals between the speaker's utterances. (This computation presently takes approximately 15 seconds on a 15-MIPS NeXT workstation.) Second, compensation of the incoming speech is expedited by normalizing only the first several cepstral coefficients rather than the entire vector, and by computing cepstral distances only for those codewords that are most similar to the incoming speech vector. The actual cepstral compensation is presently accomplished in better than real time using the Motorola 56001 DSP chip on the NeXT workstation.

Figure 6-6 shows how the recognition accuracy of the BSDCN algorithm and the real-time implementation of the CDCN algorithm depend on the amount of environment-specific speech data available for adaptation. The recognition accuracy of the real-time CDCN algorithm converges with only about 2 seconds of adapting speech, while the BSDCN algorithm requires at least 60 seconds of adapting speech to reach asymptotic levels of recognition accuracy. This is consistent with intuition, as the CDCN algorithm imposes more structure on the compensation process (from knowledge of how speech is likely to be degraded), while the BSDCN algorithm is entirely data driven.

6.5. Summary

In this chapter we have described three algorithms that are very efficient computationally: the ISDCN, the BSDCN and the FCDCN algorithm. Both ISDCN and FCDCN can be implemented with the addition

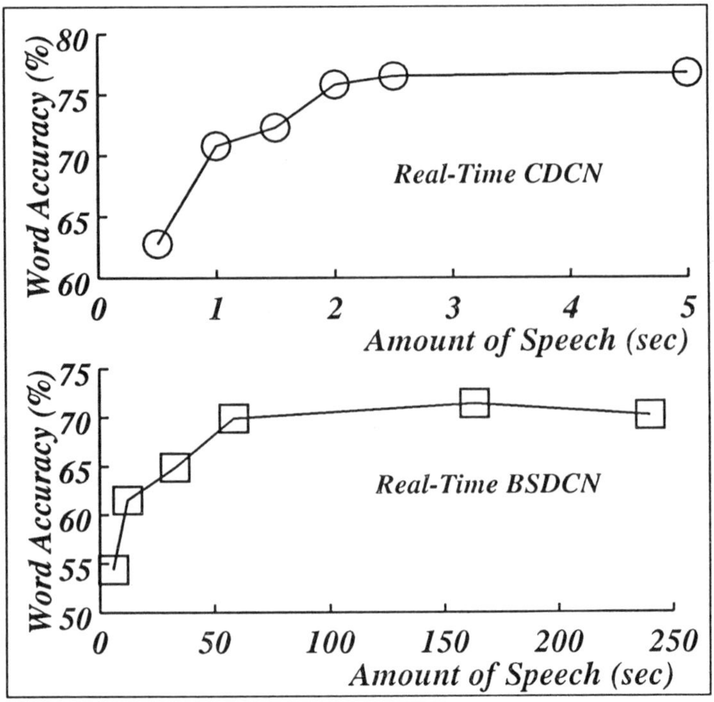

Figure 6-6: Dependence of recogntion rate of the CDCN and BSDCN algorithms on the amount of speech in the testing environment available for adaptation. The system was trained using the CLSTK microphone and tested using the PZM6FS.

of some book-keeping information at the VQ stage, whereas BSDCN is even simpler.

The Interpolated SNR-Dependent Cepstral Normalization (ISDCN) algorithm uses correction vectors that interpolate between the noise vector at low SNR to the equalization vector at high SNR. The equalization vector is estimated to minimize the accumulated VQ distortion. This is made possible by the use of an interpolating function that in our implementation was a sigmoid function.

The Blind SNR-Dependent Cepstral Normalization (BSDCN) algorithm uses compensation vectors that are a function of the instantaneous SNR exclusively, just as in SDCN. The correction vectors are estimated from long-term average cepstra at different SNRs and hence no stereo data is required.

The Fixed Codeword-Dependent Cepstral Normalization (FCDCN) algorithm uses correction vectors that depend on the instantaneous SNR of the input and that are different for every codeword. This algorithm provides a higher accuracy than the CDCN because it is free from some of the assumptions of the CDCN algorithm. On the other hand, it requires stereo training.

We showed that rapid environmental adaptation is possible with the CDCN and ISDCN algorithms (as little as 2 seconds of speech suffice), with CDCN offering a higher recognition accuracy, at the expense of a higher computational complexity. On the other hand, the simpler BSDCN requires about a minute of speech to attain its asymptotic recognition rate.

7 Frequency Normalization

In this chapter we will discuss the mel-scale cepstral parameters used in SPHINX and some optimizations that we performed on them. Also, an algorithm for frequency normalization based on variable warping of the frequency axis is presented within the framework of minimal VQ distortion. These techniques together decrease the error rate by 15 to 20%.

7.1. The Use of Mel-scale Parameters

In this section we introduce the concept of mel-scale cepstral coefficients as the DFT of the logarithm of the power spectral density function over a warped frequency scale. A rationale for its use is presented and the particular implementation of these coefficients in the SPHINX system as a matrix multiplication on LPC cepstrum coefficient is described.

The use of a warped frequency scale as opposed to a linear one has proven advantageous in speech recognition systems. Most researchers agree that frequencies above 4 kHz contribute much less to speech intelligibility and recognition accuracy than frequencies below 4 kHz. Since for systems like SPHINX the sampling rate is 16 kHz, using a linear scale would give the same weight to low and high frequencies, which is undesirable.

Further evidence for the use of a non-linear frequency axis can be extracted from the behavior of the human auditory system. Zwicker

[89] introduced the *Bark scale*[33] as an approximation to the discrimination power of different frequencies in the human auditory system.

Davis and Mermelstein [14] found that *mel-scale* coefficients resulted in improved accuracy. They used a filterbank in which the spacing among the filters and their bandwidths approximated the Bark scale. Researchers at BBN (Chow *et al.* [9]), SRI (Murveit and Weintraub [58]), MIT Lincoln Labs (Paul [68]) and some other laboratories have successfully used cepstral coefficients derived from a mel-scale filterbank approach as the front-end for their speech recognizers.

The SPHINX system also uses mel-scale cepstral coefficients as the parameter set, but they are based on an LPC analysis as opposed to filterbank. Shikano [75] applied the technique of bilinear transformation introduced by Oppenheim and Johnson [64] to warp the frequency axis. More details on the bilinear transform can be found in Appendix III.

Briefly, the bilinear transform is a mapping in the complex plane that maps the unit circle onto itself. It is defined as

$$z_{new}^{-1} = \frac{z^{-1} - \alpha}{1 - \alpha z^{-1}} \quad -1 < \alpha < 1 \tag{7.1}$$

The frequency transformation is obtained by making the substitution $z = e^{j\omega}$ and $z_{new} = e^{j\omega_{new}}$ in (7.1):

$$\omega_{new} = \omega + 2\,\mathrm{arctg}\left[\frac{\alpha \sin(\omega)}{1 - \alpha \cos(\omega)}\right] \tag{7.2}$$

Equation (7.2) is plotted in Figure 7-1 for different values of the warping parameter α.

For values of α between 0.4 and 0.8, the warping transformation is similar to the Bark scale. In the SPHINX system the parameter α was set to 0.6. Although SPHINX was not very sensitive to the particular value, 0.6 turned out to be optimum in our evaluations.

As can be seen in Appendix III, the relationship between the cepstral

[33]The *Bark scale* is approximately linear for frequencies below 1000 Hz and logarithmic above that level.

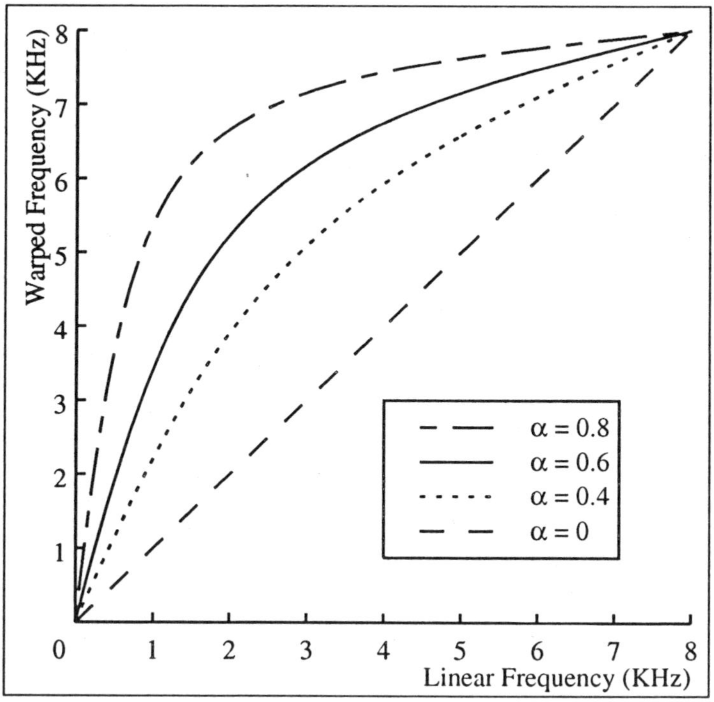

Figure 7-1: Frequency mapping of the Bilinear Transform for different values of α. Note that $\alpha=0$ is equivalent to no warping while $\alpha=0.8$ is a very severe warping of the frequency axis.

coefficients before warping $\mathbf{c}$ and after the bilinear transform $\mathbf{c}_w$ can be expressed as a matrix multiplication operation

$$\mathbf{c}_w = \mathbf{L}(\alpha)\,\mathbf{c} \tag{7.3}$$

where the warping matrix $\mathbf{L}$ is a function of α. In the original signal processing in SPHINX both $\mathbf{c}$ and $\mathbf{c}_w$ had 12 coefficients.

7.2. Improved Frequency Resolution

In this section we analyze the the frequency warping used in the standard SPHINX system and its effect on frequency resolution. Increasing the number of cepstral coefficients before the bilinear transform yielded a moderate decrease in error rate.

The truncation used in the cepstrum vector has to be done in such a way that important information is not lost. It is shown in Appendix II that the LPC-cepstrum is an infinite sequence. In Appendix III we show that the number of coefficients after the bilinear transform will be infinite. This is to say that both $\mathbf{c}$ and $\mathbf{c}_w$ in Equation (7.3) will be of infinite dimension. However, since the higher order cepstrum coefficients tend to zero rapidly, the dimension of vector $\mathbf{c}_w$ can be truncated with negligible error. In SPHINX the dimension of $\mathbf{c}_w$ was set to 12 (plus the zeroth order term).

We found that the dimension of $\mathbf{c}$ had to be larger than 12 in order not to lose frequency resolution. In Figure 7-2 we show the spectral representation of the vowel /ih/ in the word "six". We observe that the use of 12 coefficients for $\mathbf{c}$ results in a loss of resolution, that is maintained if 32 coefficients are used. With 12 cepstral coefficients, resonances whose bandwidths are smaller that $8000 / 12 = 666$ Hz will not be represented accurately because of the low-pass *liftering*. We observed that the use of 32 coefficients provided a higher low-pass *quefrency* that allowed resonances with bandwidths larger than $8000 / 32 = 250$ Hz to be retained. A dimension of 12 for $\mathbf{c}_w$ was considered sufficient because of stretching of the low frequencies, where the important resonances occur.

We trained and tested SPHINX on the census database using both 12 and 32 coefficients for $\mathbf{c}$. The results in Table 7-1 show a 7% decrease in error rate by using 32 coefficients for $\mathbf{c}$. Although the difference is not statistically significant for this task, the reduction in error rate is consistent with what was observed in the resource management task and other tasks. We considered the version with 32 coefficients as the baseline condition for all the experiments in this monograph.

At first, we were somewhat disappointed that there was only a reduction of 7% after having observed the phenomenon in Figure 7-2. The explanation that we have is that although we are not retaining additional formant information by using 32 coefficients for $\mathbf{c}$, this information is not so critical anyway if the system is speaker independent. The effect of a crisper frequency structure obtained by the use of 32 coefficients for $\mathbf{c}$ is diminished by the fact that different speakers have different formants, which would make the HMM distributions broader.

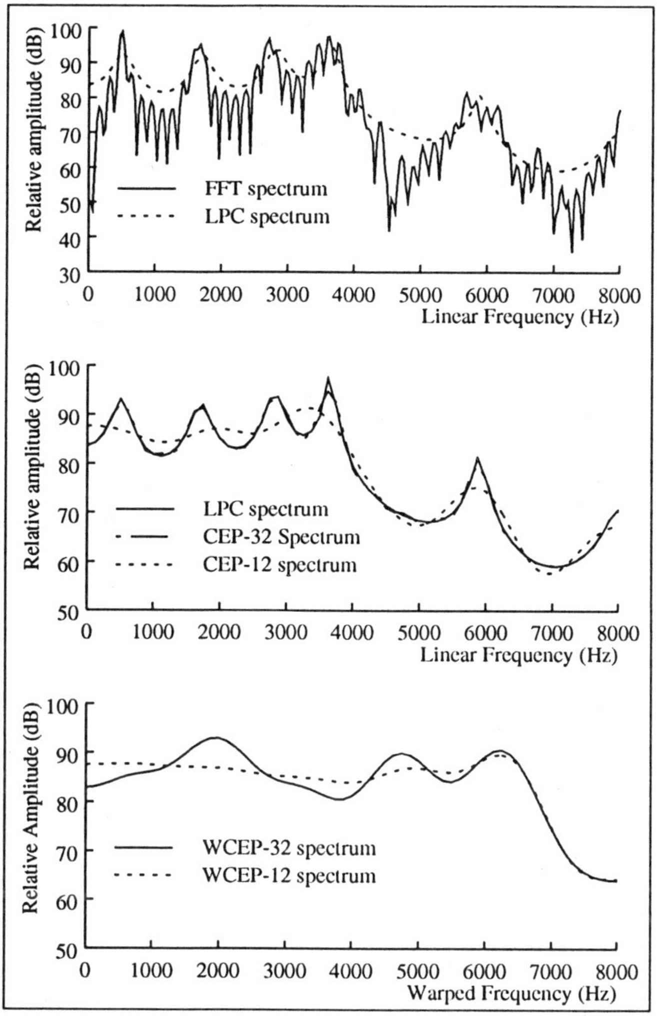

Figure 7-2: Spectral analysis of the vowel /ih/. The first graph shows the FFT spectrum and the LPC spectrum of order 14. The second graph shows the LPC spectrum and its cepstral approximation with 32 and 12 coefficients. The third graph shows the warped spectrum plotted from 12 cepstral coefficients after the bilinear transform with $\alpha = 0.6$ when both 32 and 12 coefficients are used before the transform. The use of 12 coefficients before the warping removes the formant structure.

	12 coeff	32 coeff
Recognition accuracy	84.2 %	85.3 %

Table 7-1: Effect of the use of 12 and 32 cepstral coefficients before the bilinear transform in SPHINX. The number of coefficients after frequency warping was 12 in both cases.

7.3. Variable Frequency Warping

In this section we present a novel technique for frequency normalization based on the use of the bilinear transform with a variable warping parameter. This technique is aimed at closing the gap in performance that exists between speaker-dependent and speaker-independent systems by normalizing the long-term frequency characteristics of different speakers.

Speaker-independent systems perform with an error rate that is about 3 or 4 times greater than similarly trained speaker-dependent systems (Pallett *et al.* [66]). Part of the problem can be found in that speaker-independent systems like SPHINX have to cope with the burden of different formant frequencies of different speakers, that broaden the HMM distributions. It would be desirable to be able to normalize the formant frequencies of different speakers.

It is well known that female speech exhibits higher formant frequenciesand pitch than male speech. Furthermore, a source-production model (Rabiner and Schafer [71]) suggests that the nominal resonance frequencies are essentially proportional to the length of the vocal tract. A possible mechanism for normalization would be to warp frequencies more severely for males than for females, so that after this *speaker-dependent* frequency warping, the resonance frequencies coincide.

We propose to achieve the variable frequency warping by using a different α parameter in the bilinear transform for every speaker. The value of α is selected as the one that minimizes the overall VQ distortion, the same criterion used in Chapter 6. This algorithm works in an *unsupervised* mode, since it does not require sex information or any other characterization of the speaker's formant frequencies.

For practical considerations we chose to implement the bilinear transform in two stages: the first stage being fixed and the second being variable. This is made possible due to the fact that a cascade of two bilinear transform stages of parameters α_1 and α_2 is equivalent to another bilinear transform with parameter $(\alpha_1 + \alpha_2)/(1 + \alpha_1\alpha_2)$ as shown in Appendix III. The advantage in doing this is that we use the 32 coefficients necessary for the increased frequency resolution only in the first warping stage, while we use only 12 coefficients for the second stage without any loss in accuracy.

Although we could develop a framework for maximization over continuous values of α, we choose to work with a discrete grid of values of α with $\alpha = \alpha_0 + i\Delta\alpha$ for simplicity, with $i = -N, \ldots, 0, \ldots, N$.

The new codebook is generated by the Lloyd algorithm used for finding the VQ codebook, with the difference that the α parameter for every speaker is different. This optimum α is the one that minimizes the VQ distortion for all the utterances uttered by a given speaker. Therefore only the labeling is different from the standard Lloyd algorithm.

By doing this we observed that the algorithm always found the highest value for alpha. In the degenerate case of $\alpha = 1$, every input frame is transformed into a constant (the DC value) by the bilinear transform, therefore yielding zero VQ distortion. However this trivial transformation would destroy all the information in the signal. This is the first time that we encounter that minimum distortion is not highly correlated with maximum likelihood and minimum probability of error, but quite the opposite. It was observed empirically that higher values of α_0 yielded lower distortion, although lower recognition rate.

Since we wanted to keep $\alpha_0 = 0.6$ as our center value, we had to force that constraint directly. We used 20 female speakers and 20 male speakers in training, with each speaker having his or her own α_i and we imposed that the sum of all displacements had to be zero $\Sigma\, \alpha_i = 0$ when labeling the training data.

This variable warping technique was evaluated with an optimum α per speaker. The center value α_0 was set to 0.6, with $N = 5$ and $\Delta = 0.02$. The results of the evaluation on the census database are shown in Table 7-2. As we can see, there is a 12% decrease in error rate. Use of a denser grid did not provide any additional benefit. The value of α_0 was not very critical, although 0.6 provided the maximum gain.

	BASELINE	FREQNORM
Recognition accuracy	85.3 %	87.1 %

Table 7-2: Comparison of the performance of the variable frequency warping with the baseline.

In Figure 7-3 we show the distribution of values of α_{opt} for male and female speakers. As we had anticipated there is a clear separation between them, which confirms the assumptions of the model.

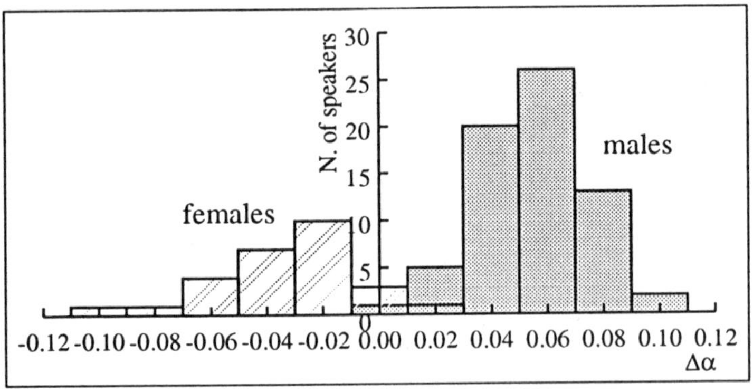

Figure 7-3: Histogram of values of α for male and female speakers.

To cross-validate the results in Table 7-2, we used the CLSTK speech from the multi-microphone database. The results of the evaluation are shown in Table 7-3. As can be observed, the algorithm never does worse than the baseline, and on the average there is a 10% decrease in error rate. There is especially a large gain in Set 1 for which the baseline performance was lower than the rest.

	BASELINE	FREQNORM
Set 1	82.4 %	85.4 %
Set 2	84.8 %	85.4 %
Set 3	87.2 %	87.8 %

Table 7-3: Comparison of the performance of the variable frequency warping with the baseline.

7.4. Summary

Mel-scale cepstrum coefficients are advantageous for speech recognition especially if a bandwidth larger than 4 kHz is used. SPHINX uses the bilinear transform as a mean of obtaining frequency-warped LPC-cepstrum.

Truncation of the cepstral coefficients is acceptable as long as they retain the frequency structure. After the bilinear transform only 12 coefficients are needed but before the transform 12 is not sufficient. We showed that the use of 32 coefficients before the bilinear transform results in no apparent loss in the accuracy of the spectral representation, and a 7% decrease in error rate.

We proposed the use of the bilinear transform with variable parameter as a means of *frequency normalization*. Histograms of the optimum α show that male and female speech are well separated, with female speech requiring a smaller α (less warping) than male speech. The use of frequency normalization provides an additional decrease in error rate of approximately 10%.

8

Summary of Results

In this chapter we summarize the results. Table 8-1 shows the performance of all the algorithms described in this monograph for the census database.

TRAIN TEST	CLSTK CLSTK	CLSTK CRPZM	CRPZM CLSTK	CRPZM CRPZM	stereo data
BASE	85.3%	18.6%	36.9%	76.5%	no
EQUAL	N/A	38.3%	50.9%	76.5%	no
PSUB	N/A	38.6%	70.6%	70.1%	no
MSUB	N/A	63.6%	71.7%	71.3%	no
MMSE1	N/A	48.7%	68.7%	71.4%	yes
EQ+MMSE1	N/A	61.4%	75.8%	74.3%	yes
EQ+MSUB	N/A	62.1%	73.7%	71.4%	yes
MMSEN	N/A	66.4%	75.5%	72.3%	yes
SDCN	N/A	67.2%	76.4%	75.5%	yes
FCDCN	N/A	73.1%	79.3%	75.8%	yes
ISDCN	84.8%	62.1%	71.4%	72.4%	no
CDCN	85.3%	74.9%	73.7%	77.9%	no

Table 8-1: Recognition accuracy of SPHINX on the census database under different normalization algorithms.

The algorithms are the following:

- BASE: Baseline case, no processing (See Chapter 2).
- EQUAL: The CRPZM speech is equalized by adding a fixed cepstral vector that is the difference between the average speech for both CLSTK and CRPZM (See Chapter 3).
- PSUB: The power spectral subtraction rule is applied to the CRPZM speech (See Chapter 3).
- MMSE1: The frequency bands in the CRPZM speech are transformed according to a curve that minimizes the squared error between CLSTK and CRPZM (See Chapter 3).
- MSUB: Magnitude spectral subtraction with under and over-subtraction. is applied to the CRPZM speech (See Chapter 3).
- EQ+MMSE1: Cascade of EQUAL and MMSE1 for the CRPZM speech (See Chapter 3).
- EQ+MSUB: Cascade of EQUAL and MSUB for the CRPZM speech (See Chapter 3).
- MMSEN: Each frequency band in the CRPZM speech is transformed with a different curve so that the squared error between the CLSTK and CRPZM is minimized (See Chapter 4).
- SDCN: The SDCN algorithm is applied to the CRPZM speech by transforming every cepstral component differently depending on the instantaneous SNR. The goal is to minimize the squared error between CLSTK and CRPZM (See Chapter 4).
- FCDCN: The FCDCN algorithm is applied to the CRPZM speech with a different correction vector for every instantaneous SNR and codeword. Again the goal is to minimize the squared error between CLSTK and CRPZM speech (See Chapter 6).
- ISDCN: The ISDCN algorithm is applied to both CLSTK and CRPZM by estimating directly the noise and equalization vector (See Chapter 6).
- CDCN: The CDCN algorithm is applied to both CLSTK and CRPZM by estimating directly the noise and equalization vector (See Chapter 5).

Since we are mostly concerned with the performance when the system is trained with the CLSTK speech, we show in Figures 8-1, 8-2 and 8-3 the word accuracy of the system trained on the CLSTK and tested on the

CRPZM speech for the case of algorithms that attempt independent compensation for noise and spectral tilt, algorithms that do independent versus joint compensation and algorithms that operate in the spectral versus the cepstral domain.

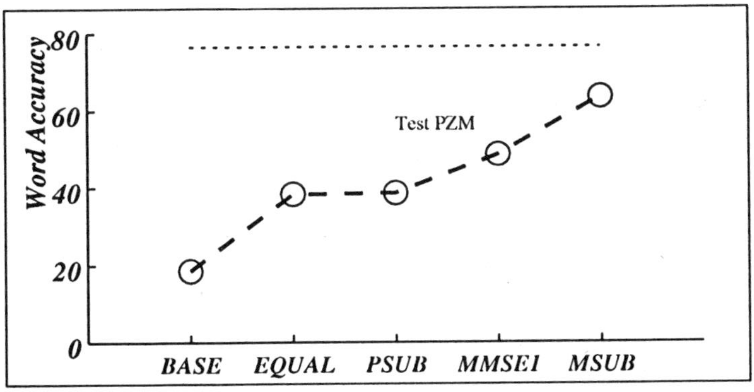

Figure 8-1: Independent compensation for noise and filtering in the spectral domain. Comparison of the baseline performance (BASE), cepstral equalization (EQUAL), Power Spectral Subtraction (PSUB), Minimum Mean Squared Error with one curve for all frequencies (MMSE1) and Magnitude Spectral Subtraction (MSUB) when trained on the CLSTK microphone and tested on the CRPZM. The broken line represents the word accuracy of the system trained and tested on the CRPZM.

Figure 8-1 shows different compensation schemes in the frequency domain that combat the linear filtering (EQUAL) or additive noise (PSUB, MMSE1, MSUB). MSUB is the one with highest accuracy of all of them. However, it is still far from the accuracy obtained when the system is trained and tested on the CRPZM microphone.

Figure 8-2 shows a comparison of the performance of algorithms that attempt an independent compensation (cascade of equalization EQUAL and noise suppression MMSE1 and MSUB) and a joint compensation for noise and filtering (MMSEN) in the frequency domain. We note that the MMSEN algorithm, by using a different transformation curve per frequency, deals better with the colored noise present in the CRPZM recordings. Also, the

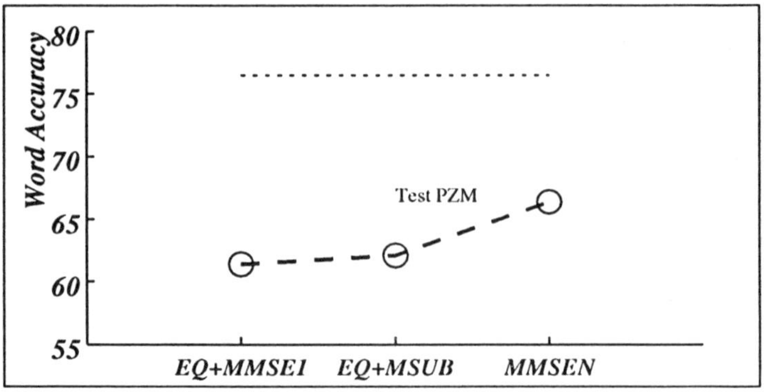

Figure 8-2: Comparison of algorithms that perform independent compensation for both noise and filtering (EQ+MMSE1, EQ+MSUB) versus an algorithm that performs joint compensation for noise and filtering (MMSEN) in the spectral domain. The Figure shows the word accuracy of SPHINX when trained on the CLSTK microphone and tested on the CRPZM. The broken line represents the word accuracy of the system trained and tested on the CRPZM.

performance of the MMSEN algorithm is higher than that of the cascade of equalization EQUAL and noise suppression MMSE1, with this difference being statistically significant.

Figure 8-3 shows a comparison of our best algorithm operating in the frequency domain (MMSEN) and the algorithms derived in the cepstral domain (SDCN, FCDCN). All three algorithms perform joint compensation for noise and filtering. The SDCN algorithm performs at the same level than the MMSEN algorithm but it is simpler computationally, as it only requires to compensate the first two cepstral components. The FCDCN on the other hand performs substantially better than both the MMSEN and the SDCN algorithms as the compensation vector is *codeword-dependent.*

Figure 8-4 shows the performance of ISDCN and CDCN as algorithms that adapt to new acoustical environments. Both of them operate in the cepstral domain and perform joint compensation for noise and filtering. As can be seen, use of the CDCN algorithm brings the accuracy of the system when trained on the CLSTK and tested on the CRPZM to the level obtained

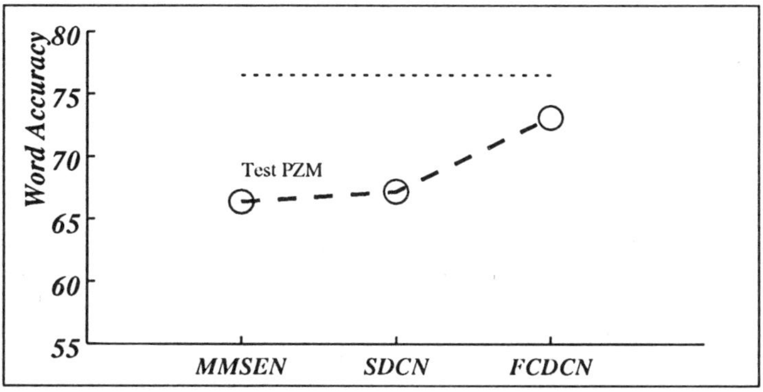

Figure 8-3: Comparison of recognition accuracy of algorithms in the spectral domain (MMSEN) and the cepstral domain (SDCN,FCDCN). All these algorithms perform joint compensation for noise and filtering but they also require stereo data. MMSEN uses a minimum mean squared error criterion with one transformation curve per frequency component, and it is described in Chapter 4. SDCN, SNR-Dependent Cepstral Normalization , and the FCDCN, Fixed Codeword-Dependent Cepstral Normalization are described in Chapters 4 and 6 respectively. SPHINX was trained on the CLSTK microphone and tested on the CRPZM. The broken line represents the word accuracy of the system trained and tested on the CRPZM.

when the system is trained and tested on the CRPZM.

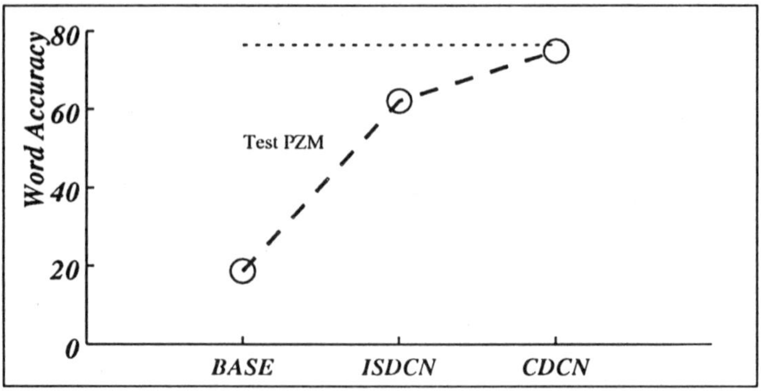

Figure 8-4: Comparison of algorithms that adapt to new acoustical environments (ISDCN and CDCN) algorithms when trained on the CLSTK microphone and tested on the CRPZM. BASE represents no processing, ISDCN is the Interpolated SNR-Dependent Cepstral Normalization described in Chapter 6 and CDCN is the Codeword-Dependent Cepstral Normalization described in Chapter 5. ISDCN and CDCN perform also joint compensation for noise and filtering. The broken line represents the word accuracy of the system trained and tested on the CRPZM.

9 Conclusions

In this monograph we have addressed the problem of building speech recognition systems that are robust to changes in the acoustical environment. With the development of large-vocabulary continuous-speech speaker-independent recognition systems like SPHINX, three of the major problems in speech recognition have been solved. However, for unconstrained speech recognition we need to build systems that will work for ungrammatical spontaneous speech, speakers with different dialects, and in real acoustical environments.

Not only have we learned many lessons through the course of this work about the problems in environment independence, but we have also provided some solutions that can be used in practice.

9.1. Contributions

The major contribution of this work is to show that an increased robustness to changes in the environment can be achieved. Use of our algorithms on a system trained on clean speech brings the performance when testing on a given environment to at least the level obtained when the system was trained and tested on that particular environment, and this can be accomplished without the need for retraining on data from the new environment. When a signal is corrupted by noise some information is inevitably lost. While we have not overcome the degradation present in noisy speech in our work, we believe that the algorithms described provide a step in the right direction.

We have proposed specific algorithms that adapt to new acoustical environments without the need for retraining. The CDCN, *Codeword-Dependent Cepstral Normalization* estimates the noise and equalization cepstral vectors that maximizes the probability of an ensemble of input cepstral frames. The ISDCN, *Interpolated SNR-Dependent Cepstral Normalization* estimates the environmental parameters by minimizing the accumulated distortion in the vector quantizer. The BSDCN, *Blind SNR-Dependent Cepstral Normalization*, overcomes the lack of stereo training by lumping together all cepstral vectors within the same SNR. With CDCN, the recognition accuracy when testing on speech recorded with a desk-mounted microphone on a system trained on speech recorded with a headset-mounted microphone is essentially the same obtained when the system is trained and tested with the recordings from the desk-mounted microphone.

We have shown that *joint compensation* for noise and spectral equalization is more effective than a combination of independent compensators. The correction vectors are an interpolating function between the equalization vector at high SNR and the noise vector at low SNR. Conditioning on the *instantaneous SNR* can provide processing benefits while retaining simplicity. We have identified instantaneous SNR as the primary variable in the normalization procedure.

The use of a *universal codebook* is the main tool that allows us to track deviations from a standard acoustical ambience by finding the parameters of the transformation that best "matches" different acoustic spaces. Therefore, a universal codebook that is trained once and for all can be used instead of training different codebooks for different applications. Vocabulary-independent models (Hon and Lee [36]) will benefit from this approach, as those models need to be environment-independent as well.

We have shown that the *cepstral domain* is a viable parameter domain for noise suppression. The advantage of performing normalization in the cepstral domain is that we can integrate it better with the rest of a system like SPHINX that uses cepstral parameters as feature vectors.

The use of the *EM algorithm* as a tool for obtaining maximum likelihood estimates has been essential to the robust estimation of parameters used in this monograph.

We have shown that ISDCN and FCDCN can be implemented by modifying the VQ stage, resulting in an algorithm that is very *efficient* computationally.

We have introduced the use of *stereo databases* for evaluation purposes. While it is difficult to compare the performance of algorithms when there is not a standard database that all researchers use, some criteria can be established that will help to make comparisons. We proposed the use of two reference figures: the error rate when the system is trained and tested on clean speech and the error rate when the system is trained and tested on the noisy speech. These references will serve as indicators of the performance of our algorithms when the system is trained on clean speech and tested on noisy speech.

We have also introduced the use of *speech and noise spectral averages* as a more informative feature than just SNR as a characterization of an environment. When speech can be passed through a linear filter and the noise is not white, SNR measurements can provide a misleading indication of how a speech recognizer will perform if no normalization is accomplished.

We have proposed a method for *frequency normalization* via variable warping of the frequency axis. Although this is something that many researchers have attempted, it is not an easy task because the frequency alignment is very unconstrained. The use of the bilinear transform with only one free parameter has proven to be effective in normalizing the the nominal resonance frequencies of different speakers.

9.2. Suggestions for Future Work

Even though we have taken a big step forward in being able to transfer speech recognition systems from the laboratory to the real world, the field of environment-independent recognition is still in its infancy. In this section we describe some future research that appears to be worthwhile.

One of the problems with the ISDCN and CDCN algorithms is that obtaining estimates of the equalization vector via maximum likelihood does not use any *a priori* information that may be available about the environment. Although the estimates are unbiased, they may exhibit a large variance if not enough samples are used. We have observed that in some cases, these estimates may not represent legitimate equalization vectors. A topic to investigate in future research is the use of *a priori* information on the distribution of the noise and equalization vectors. One possibility would be to include a small codebook of different environments. The highly accurate FCDCN algorithm could be used with different sets of correction vectors depending on which environment is selected. As always, the criterion could still be minimization of the VQ distortion. Interpolation between different environments is also a possibility to explore.

The use of a different criterion for the estimation of the environmental parameters should also be considered in future research. In this monograph we have used the minimization of the VQ distortion or ML estimation at the acoustic level as the criterion to estimate the environmental parameters. It may well be the case, however, that a criterion such as maximization of the probability of the utterance or minimization of the probability of error could lead to better performance. Although minimization of the probability of error would be the goal of any speech recognition system, it would be extremely difficult to use as the basis for a distortion metric. Maximization of the probability of the utterance, however, is a plausible alternative and it is the criterion used by HMMs. In other words, we could leave the estimation of environmental parameters to the HMM search, so that we are not penalized by early decisions. Although there could be a small improvement in performance if this is done at the VQ stage by using tied mixtures or semi-continuous HMM models (Huang *et al.* [37]), we believe that most of the benefit would come from a more accurate estimation of the environmental parameters.

We have found that all algorithms described in this monograph can potentially perform rather well for high SNRs, and that the problems arise when frames with low SNR are processed. For those frames, the algorithms select just one "cleaned" vector even though there will be several possible that have been masked by the noise. Selecting one of them and not including the other alternatives is a serious flaw of the present approaches, as it would be better to let the HMM search makes its decision with all information available. One way of doing this would be rather than using discrete models to use semi-continuous ones in which the relative probabilities of different codewords are passed to the HMM search. Another possible alternative would be to use some sort *state-dependent* (like in [59], [84])rather than *codeword-dependent* correction vectors. Specifically, we propose the use of different sets of correction vectors per state of an HMM (we could cluster them to retain only a handful of state-dependent corrections). Many different correction vectors and VQ labels would be obtained for every frame of speech depending on what phone was hypothesized. The HMM search would select the most likely phone string.

A longer time frame can help in the compensation process. In the work in this monograph, compensation was based only on knowledge of the current frame, while it is clear that using information on the adjacent frames will benefit the accuracy of the estimate. A possible extension of the techniques developed here would be to incorporate the same processing to differential parameters.

The algorithms described in this monograph could also be used for speech enhancement. The CDCN algorithm, for example, could be used to obtain a better spectral estimate of the clean speech than that of conventional noise suppression algorithms. Wiener filtering could then be used to enhance the signal.

More work remains to be done on the problem of real-time implementation. The algorithms described in this monograph aim to normalize long-term characteristics of the speech signal in addition to the short-term ones used by most systems. Allowing for a slowly changing environment requires more research on the rate of change and the mechanisms for updating. In this work we have shown that using the previous utterance as a unit is a reasonable choice (or about 2 seconds of speech), but that the use of longer estimation times will yield more accurate estimates.

This monograph has not explicitly addressed the problem of interference by additive non-stationary noise. We are optimistic that the technique of noise-word modeling described by (Ward [85]) will provide some additional improvement for speech collected in the presence of non-stationary interference sources such as slamming doors, ringing telephones, etc. This use of noise-word models is a complementary technique to the algorithms described in this monograph.

It is necessary to investigate the behavior of many different microphones and environments, so that more general conclusions can be drawn. It would also be desirable to test these algorithms with telephone speech. In particular it would be extremely helpful to test these algorithms on non-linear devices used in communications such as carbon microphones. We believe that the data-driven algorithms described in this monograph (SDCN, ISDCN, BSDCN, FCDCN) could recover from the distortion better than algorithms that have a clear model for the speech degradation like CDCN.

The absence of a standard database for evaluation of algorithms is definitely not benefiting the field of environment-independent recognition. The continuous speech recognition community has benefited from the existence of common tasks and databases so that direct comparison between algorithms is more straightforward. Our hope is that in the near future similar efforts can be directed toward the development of standard databases for environment-independent recognition.

Appendix I
Glossary

I.1. Time Domain

$x[m]$	The clean signal
$y[m]$	The noisy signal
$h[m]$	The filter's impulse response
$n[m]$	The noise process

I.2. Frequency Domain

$X_i(\omega)$	Power Spectral Density of clean signal at frame i
$Y_i(\omega)$	Power Spectral Density of noisy signal at frame i
$N(\omega)$	Power Spectral Density of the noise
$Q(\omega) = \lvert H(\omega) \rvert^2$	Magnitude squared of the filter's frequency response
$\mathbf{X}_i(\omega) = \ln X_i(\omega)$	Log-spectrum of clean signal at frame i
$\mathbf{Y}_i(\omega) = \ln Y_i(\omega)$	Log-spectrum of noisy signal at frame i
$\mathbf{N}(\omega) = \ln N(\omega)$	Log-spectrum of the noise
$\mathbf{Q}(\omega) = \ln Q(\omega)$	Equalization transfer function
$\overline{\mathbf{X}}_i(\omega) = \mathbf{X}_i(\omega) - \mathbf{N}_i(\omega)$	Normalized log-spectrum of clean signal at frame i
$\overline{\mathbf{Y}}_i(\omega) = \mathbf{Y}_i(\omega) - \mathbf{N}_i(\omega)$	Normalized log-spectrum of noisy signal at frame i

I.3. Cepstral Domain

$\mathbf{x}_i$	Cepstral vector of clean signal at frame i
$\mathbf{y}_i$	Cepstral vector of noisy signal at frame i
$\mathbf{z}_i$	Observed cepstral vector at frame i
Z	Collection of cepstral vectors for an utterance
$\mathbf{n}$	Cepstral vector of the noise
$\mathbf{q}$	Cepstral equalization vector
$\mathbf{r}[\mathrm{k}]$	Correction vector $\mathbf{r}$ for the kth codeword
$\mathbf{s}[\mathrm{k}]$	Correction vector $\mathbf{s}$ for the kth codeword
$\mathbf{w}(\mathrm{SNR})$	Correction vector $\mathbf{w}$ as a function of SNR
$\mathbf{c}[\mathrm{k}]$	Vector of the kth codeword

I.4. Indices

The following indices refer to

i	Frames in an utterance
k	Codewords in codebook
j	Iteration number in iterative algorithms

I.5. Probabilistic Models

S	The word string
ν	The noise model
ξ	The speech model
$P[k]$	*A priori* probability for k^{th} mixture component
Σ_k	Covariance matrix of kth mixture component

Appendix II
Signal Processing in Sphinx

The speech signal $x[m]$ is digitized at a sampling rate of 16 kHz and multiplied by a Hamming window $h[m]$ of $N = 320$ samples (20 ms) every $M = 160$ samples (10 ms).

$$h[m] = 0.54 - 0.46\cos\left(\frac{2\pi m}{N-1}\right) \quad 0 \le m \le N-1 \tag{II.1}$$

$$x_i[m] = x[iM + m]\,h[m] \quad 0 \le m \le N-1 \tag{II.2}$$

where $x_i[m]$ represents frame i. After this, a high-pass preemphasis filter is applied

$$y_i[m] = x_i[m] - 0.97\,x_i[m-1] \tag{II.3}$$

The preemphasized windowed signal $y_i[m]$ is used to compute $p = 14$ autocorrelation coefficients

$$R[k] = \sum_{m=0}^{N-1-k} y_i[m]\,y_i[m+k] \quad 0 \le k \le p \tag{II.4}$$

that are multiplied by a Pascal lag window with $\tau = 1500$

$$v[k] = \frac{\binom{\tau-k+1}{k}}{\binom{\tau+k-1}{k}} \tag{II.5}$$

that has the form in Figure II-1.

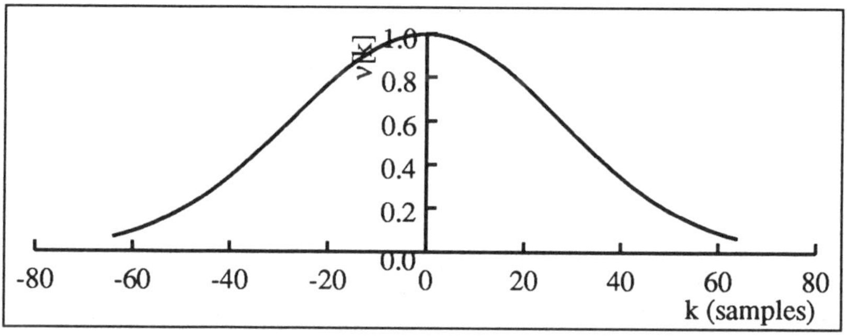

Figure II-1: Pascal window for $\tau = 1500$

The lag window will suppress the harmonics that will appear at multiples of the pitch period, which can be as low as 32 samples at 16 kHz for a high pitched female voice with a fundamental frequency of 500 Hz. The Levinson-Durbin recursion (it can be found in Rabiner and Schafer [71] for example) is used to obtain the LPC parameters from the windowed autocorrelation coefficients

$$E^{(0)} = R[0] \tag{II.6}$$

$$k_i = \frac{R[i] - \sum_{j=1}^{i-1} a_j^{(i-1)} R[i-j]}{E^{(i-1)}} \quad 1 \le i \le p \tag{II.7}$$

$$a_i^{(i)} = k_i \tag{II.8}$$

$$a_j^{(i)} = a_j^{(i-1)} - k_i a_{i-j}^{(i-1)} \quad 1 \le j \le i-1 \tag{II.9}$$

$$E^{(i)} = (1 - k_i^2) E^{(i-1)} \tag{II.10}$$

Equations (II.7) - (II.10) are solved recursively for $i = 1, 2, \ldots, p$ with the final solution being given by

$$a_j = a_j^{(p)} \quad 1 \le j \le p \tag{II.11}$$

The LPC parameters define an all-pole system

$$H(z) = \frac{G}{1 - \sum_{k=1}^{p} a_k z^{-k}} \tag{II.12}$$

whose logarithm can be expressed as a Fourier series with the *complex* cepstrum. Since it is obtained from an LPC analysis, this is called the LPC-cepstrum.

$$\ln H(z) = \sum_{n=0}^{\infty} \mathbf{c}[n] z^{-n} \tag{II.13}$$

Atal [3] showed that taking the logarithm of (II.12), equating it to (II.13) and taking derivatives with respect to z^{-1} leads to the following recursion for $c[n]$

$$\mathbf{c}[0] = \ln(G)$$

$$\mathbf{c}[n] = a_n + \sum_{k=1}^{n-1} \left(\frac{k}{n}\right) \mathbf{c}[k] \, a_{n-k} \quad 1 \le n \tag{II.14}$$

Even though there is an infinite number of cepstrum coefficients, we truncated the sequence to 32. Finally a bilinear transform with warping parameter $\alpha = 0.6$ is applied as described in Appendix III, obtaining 12 frequency-warped LPC-cepstral coefficients.

Appendix III
The Bilinear Transform

The transformation defined by

$$z_{new}^{-1} = \frac{z^{-1} - \alpha}{1 - \alpha z^{-1}} \qquad (III.1)$$

for $0 < \alpha < 1$ belongs to the class of so-called bilinear transforms. It is a mapping in the complex planethat maps the unit circle onto itself. The frequency transformation is obtained by making the substitution $z = e^{j\omega}$ and $z_{new} = e^{j\omega_{new}}$:

$$\omega_{new} = \omega + 2\,\mathrm{arctg}\left[\frac{\alpha \sin(\omega)}{1 - \alpha \cos(\omega)}\right] \qquad (III.2)$$

An algorithm to perform this frequency warping transformation on a time signal is described by Oppenheim and Johnson [64]. If we assume a causal finite time sequence $c[n]$, with $c[n] = 0$ for $n < 0$ and $n > p$, the warped sequence $c_w[n]$ will contain an *infinite* number of coefficients that can be obtained by passing the reversed input sequence through a cascade of filters and sampling their outputs at time $n = 0$ as shown in Figure III-1.

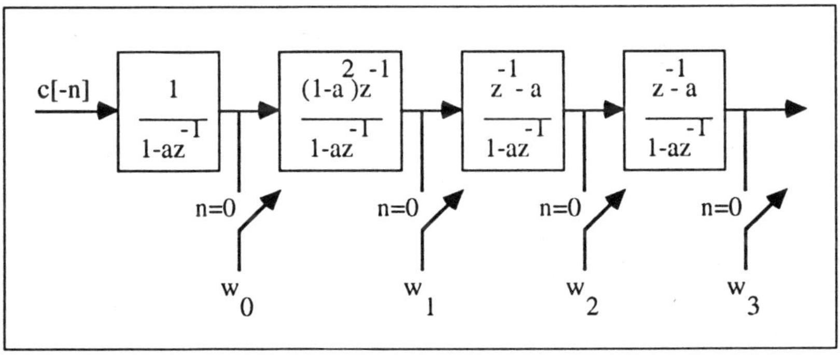

Figure III-1: Bilinear transform algorithm as a linear filtering operation. By having as the input to this stage of filters a time-reversed cepstrum sequence, we can obtain the corresponding warped coefficients as the outputs of these filters at time $n = 0$.

This algorithm is used by SPHINX to do the warping transformation on the cepstral coefficients. We have to use the *complex* cepstrum rather than the *real* cepstrum[34] because the sequence $\mathbf{c}[n]$ has to be causal to use Oppenheim's algorithm.

We have noticed that if only p coefficients are retained for $\mathbf{c}_w[n]$, we can relate the input/output relationship by a linear transformation of the form

$$\mathbf{c}_w = \mathbf{L}(\alpha)\,\mathbf{c} \qquad (III.3)$$

where $\mathbf{c}$ and $\mathbf{c}_w$ are the input and output vectors of length p and $\mathbf{L}$ is the warping matrix. Nocerino *et al.* [62] also pointed out that a general warping transformation is equivalent to a matrix multiplication. We see that every coefficient in the warped sequence $\mathbf{c}_w[n]$ is a linear combination of the coefficients of the original sequence $\mathbf{c}[n]$. If the warping parameter α is small, it can be shown that neglecting all powers of α greater than 1, the finite length sequence $\{\mathbf{c}[n], n=0, 1, \ldots, p\}$ is transformed as

$$\mathbf{c}_w[n] = -(n-1)\,\alpha\,\mathbf{c}[n-1] + \mathbf{c}[n] + (n+1)\,\alpha\,\mathbf{c}[n+1] \qquad (III.4)$$

III.1. Cascade of Warping Stages

We now show that the application of two stages of the bilinear transform with warping parameters α_1 and α_2 is equivalent to applying one stage of the bilinear transform with $\alpha = (\alpha_1 + \alpha_2)/(1 + \alpha_1\alpha_2)$.

Let z be the complex variable in the original and s and u the complex variable after one and two bilinear transforms. The relationship between them is

$$s^{-1} = \frac{z^{-1} - \alpha_1}{1 - \alpha_1 z^{-1}} \qquad (III.5)$$

$$u^{-1} = \frac{s^{-1} - \alpha_2}{1 - \alpha_2 s^{-1}} \qquad (III.6)$$

Combining (III.5) and (III.6) we obtain

[34]The *real* cepstrum is the even part of the *complex* cepstrum. The interested reader is referred to Rabiner and Schafer [71] for details.

$$u^{-1} = \frac{\dfrac{z^{-1} - \alpha_1}{1 - \alpha_1 z^{-1}} - \alpha_2}{1 - \alpha_2 \dfrac{z^{-1} - \alpha_1}{1 - \alpha_1 z^{-1}}}$$

$$= \frac{z^{-1}(1 + \alpha_1\alpha_2) - (\alpha_1 + \alpha_2)}{(1 + \alpha_1\alpha_2) - (\alpha_1 + \alpha_2)z^{-1}} = \frac{z^{-1} - \alpha}{1 - \alpha z^{-1}} \qquad (III.7)$$

where the new α can be expressed as

$$\alpha = \frac{\alpha_1 + \alpha_2}{1 + \alpha_1\alpha_2} \qquad (III.8)$$

Appendix IV
Spectral Estimation Issues

In this appendix we discuss the characterization of the LPC-cepstrum spectral estimator as a zero-mean Gaussian random vector. The use of a finite data sample and possible inadequacies of the model will not make possible to obtain the exact PSD (Power Spectral Density) of the speech signal.

The PSD of a random process cannot be inferredfrom a sample function of the process. We know that if the random process is ergodic, an infinite sample function has all the information required to obtain the PSD of that process. Since speech is a non-stationary random process, the spectral estimation techniques used are based on quasi-stationarity or short-term analysis. The data window used in SPHINX is 320 points, which is short enough as to assume stationarity and long enough as to yield reasonable estimates.

The statistics of the LPCparameters and the AR Power Spectral Density Estimator are not available even for an analysis of order 1. Kay [44] showed that the spectrum of the LPC estimator is asymptotically Gaussian (when both the number of points and the analysis order tend to infinity). Furthermore, the log-spectral estimate has a variance independent of the mean value.

By using an LPC analysis, we have made the assumption that the speech signal can be characterized as an AR process. Since for instance, nasals present zeroes as well as poles, there will be some inadequacies by using the LPC-cepstrum.

We assume that the frequency-warped LPC-cepstrum $\mathbf{z}$ computed by the SPHINX front-end is a noisy estimate of the true *Power Cepstral Density* $\mathbf{y}$. In the absence of exact statistics we modeled the *pdf* $p(\mathbf{z}/\mathbf{y})$ as a Gaussian random vector $N_z(\mathbf{y},\Gamma)$. We have confirmed the validity of the Gaussian assumption empirically for the frequency-warped LPC-cepstrum in SPHINX. The assumption we made in this work is that the cepstral distance is normally distributed with a standard deviation γ. For the front-end in SPHINX this value resulted to be approximately 0.3, which was obtained by passing to the front-end a long sequence of artificially generated white noise, and by computing the variance of noise segments across the census database.

Appendix V
MMSE Estimation in the CDCN Algorithm

In this section we derive the expressions for the conditional MMSE estimate $E[\mathbf{x} \mid \mathbf{z}, \mathbf{n}, \mathbf{q}]$, as well as a Gaussian decomposition for the densities involved in the derivation of the CDCN algorithm.

V.1. The Conditional Probability in the CDCN Algorithm

The conditional probability of the clean vector **x** given the observation **z** and the environmental parameters **n** and **q** can be expressed as:

$$p(\mathbf{x} \mid \mathbf{z}, \mathbf{n}, \mathbf{q}) = \sum_{k=0}^{K-1} p(\mathbf{x}, k \mid \mathbf{z}, \mathbf{n}, \mathbf{q}) \tag{V.1}$$

where each term in the sum is according to Bayes Rule

$$p(\mathbf{x}, k \mid \mathbf{z}, \mathbf{n}, \mathbf{q}) = \frac{p(\mathbf{z}, \mathbf{x}, k \mid \mathbf{n}, \mathbf{q})}{p(\mathbf{z} \mid \mathbf{n}, \mathbf{q})} \tag{V.2}$$

whose numerator can be expressed as

$$p(\mathbf{z}, \mathbf{x}, k \mid \mathbf{n}, \mathbf{q}) = p(\mathbf{z} \mid \mathbf{x}, k, \mathbf{n}, \mathbf{q})\, p(\mathbf{x}, k \mid \mathbf{n}, \mathbf{q}) \tag{V.3}$$

Since the distribution of **x** does not depend on the noise nor the filter

$$p(\mathbf{x}, k \mid \mathbf{n}, \mathbf{q}) = p(\mathbf{x}, k) = P[k]\, p(\mathbf{x} \mid k) \tag{V.4}$$

where $p(\mathbf{x} \mid k)$ is the *pdf* for mixture k and $P[k]$ is the *a priori* probability for mixture k. Combining (V.3) and (V.4), the numerator in (V.2) can be expressed as

$$p(\mathbf{z}, \mathbf{x}, k \mid \mathbf{n}, \mathbf{q}) = P[k]\, p(\mathbf{z} \mid \mathbf{x}, k, \mathbf{n}, \mathbf{q})\, p(\mathbf{x} \mid k) \tag{V.5}$$

with the denominator in (V.2) having the form

$$\begin{aligned} p(\mathbf{z} \mid \mathbf{n}, \mathbf{q}) &= \sum_{k=0}^{K-1} \int p(\mathbf{z}, \mathbf{x}, k \mid \mathbf{n}, \mathbf{q})\, d\mathbf{x} \\ &= \sum_{k=0}^{K-1} \int P[k]\, p(\mathbf{z} \mid \mathbf{x}, k, \mathbf{n}, \mathbf{q})\, p(\mathbf{x} \mid k)\, d\mathbf{x} \end{aligned} \tag{V.6}$$

Combining Equations (V.1), (V.2), (V.5) and (V.6) we obtain the following expression for the *a posteriori* probability:

$$p(\mathbf{x}|\mathbf{z},\mathbf{n},\mathbf{q}) = \frac{\sum_{k=0}^{K-1} P[k]\, p(\mathbf{z}|\mathbf{x},\mathbf{n},\mathbf{q},k)\, p(\mathbf{x}|k)}{\sum_{k=0}^{K-1} P[k] \int p(\mathbf{z}|\mathbf{x},\mathbf{n},\mathbf{q},k)\, p(\mathbf{x}|k)\, d\mathbf{x}} \qquad (V.7)$$

With the *a posteriori* probability (V.7) we can derive the MMSE estimate for $\mathbf{x}$ as:

$$\hat{\mathbf{x}}_{MMSE} = E[\mathbf{x}|\mathbf{z},\mathbf{n},\mathbf{q}] = \frac{\sum_{k=0}^{K-1} P[k] \int \mathbf{x}\, p(\mathbf{z}|\mathbf{x},\mathbf{n},\mathbf{q},k)\, p(\mathbf{x}|k)\, d\mathbf{x}}{\sum_{k=0}^{K-1} P[k] \int p(\mathbf{z}|\mathbf{x},\mathbf{n},\mathbf{q},k)\, p(\mathbf{x}|k)\, d\mathbf{x}} \qquad (V.8)$$

where $p(\mathbf{x}|k)$ is the kth mixture component and $p(\mathbf{z}|\mathbf{x},k,\mathbf{n},\mathbf{q})$ is the the *pdf* of the spectral estimator $p(\mathbf{z}|\mathbf{y})$, both assumed to be Gaussian random vectors.

V.2. Gaussian Decomposition in CDCN

In this section we derive a decomposition for the product of Gaussians

$$p(\mathbf{z}|\mathbf{x},\mathbf{n},\mathbf{q},k)\, p(\mathbf{x}|k) = \frac{1}{\alpha|\Sigma_k|^{1/2}} exp[-\frac{1}{2}(\mathbf{x}-\mathbf{c}[k])^T \Sigma_k^{-1} (\mathbf{x}-\mathbf{c}[k])]$$
$$\frac{1}{\alpha'|\Gamma|^{1/2}} exp[-\frac{1}{2}(\mathbf{z}-\mathbf{x}-\mathbf{q}-\mathbf{r}(\mathbf{x}))^T \Gamma^{-1} (\mathbf{z}-\mathbf{x}-\mathbf{q}-\mathbf{r}(\mathbf{x}))] \qquad (V.9)$$

that is more convenient for integration of (V.8). By grouping the terms that depend on $\mathbf{x}$ in the exponent of (V.9) we obtain an alternate expression in terms of vector $\mathbf{b}(\mathbf{x},k)$ and the scalar $d(\mathbf{x},k)$

$$(\mathbf{z}-\mathbf{x}-\mathbf{q}-\mathbf{r}(\mathbf{x}))^T \Gamma^{-1} (\mathbf{z}-\mathbf{x}-\mathbf{q}-\mathbf{r}(\mathbf{x})) + (\mathbf{x}-\mathbf{c}[k])^T \Sigma_k^{-1} (\mathbf{x}-\mathbf{c}[k])$$
$$= (\mathbf{x}-\mathbf{b}(\mathbf{x},k))^T (\Gamma^{-1} + \Sigma_k^{-1}) (\mathbf{x}-\mathbf{b}(\mathbf{x},k)) + d(\mathbf{x},k) \qquad (V.10)$$

By equating the terms in $\mathbf{x}$ in (V.10) we obtain

$$\Gamma^{-1} (\mathbf{z}-\mathbf{q}-\mathbf{r}(\mathbf{x})) + \Sigma_k^{-1} \mathbf{c}[k] = (\Gamma^{-1} + \Sigma_k^{-1})\, \mathbf{b}(\mathbf{x},k) \qquad (V.11)$$

Also equating the zero order terms in (V.10), we get an expression for $d(\mathbf{x},k)$ of the form

$$d(\mathbf{x},k) = (\mathbf{z}-\mathbf{q}-\mathbf{r}(\mathbf{x}))^T \Gamma^{-1} (\mathbf{z}-\mathbf{q}-\mathbf{r}(\mathbf{x}))$$
$$+ \mathbf{c}[k] \Sigma_k^{-1} \mathbf{c}[k] - \mathbf{b}(\mathbf{x},k)^T (\Gamma^{-1} + \Sigma_k^{-1})\, \mathbf{b}(\mathbf{x},k) \qquad (V.12)$$

We will now try to obtain closed-form expressions for $\mathbf{b}(\mathbf{x}, k)$ and $d(\mathbf{x}, k)$ by using the following matrix identities:

$$(\Gamma^{-1} + \Sigma_k^{-1})^{-1}\,\Gamma^{-1} = (\Gamma(\Gamma^{-1} + \Sigma_k^{-1}))^{-1}$$
$$= (I + \Gamma\Sigma_k^{-1})^{-1} = ((\Sigma_k + \Gamma)\,\Sigma_k^{-1})^{-1} = \Sigma_k(\Sigma_k + \Gamma)^{-1} \qquad (V.13)$$

and similarly

$$(\Gamma^{-1} + \Sigma_k^{-1})^{-1}\,\Sigma_k^{-1} = \Gamma(\Sigma_k + \Gamma)^{-1} \qquad (V.14)$$

We get an expression for $\mathbf{b}(\mathbf{x}, k)$ by combining (V.11), (V.13) and (V.14):

$$\mathbf{b}(\mathbf{x}, k) = \Sigma_k(\Sigma_k + \Gamma)^{-1}(\mathbf{z} - \mathbf{q} - \mathbf{r}(\mathbf{x})) + \Gamma(\Sigma_k + \Gamma)^{-1}\,\mathbf{c}[k] \qquad (V.15)$$

Combining (V.11) and (V.15), we obtain

$$\mathbf{b}(\mathbf{x}, k)^T(\Gamma^{-1} + \Sigma_k^{-1})\,\mathbf{b}(\mathbf{x}, k)$$
$$= (\mathbf{z} - \mathbf{q} - \mathbf{r}(\mathbf{x}))^T\Gamma^{-1}\Sigma_k(\Gamma + \Sigma_k)^{-1}(\mathbf{z} - \mathbf{q} - \mathbf{r}(\mathbf{x}))$$
$$+ \mathbf{c}^T[k]\,\Sigma_k^{-1}\Gamma(\Sigma_k + \Gamma)^{-1}\mathbf{c}[k] + 2\,\mathbf{c}^T[k]\,(\Sigma_k + \Gamma)^{-1}(\mathbf{z} - \mathbf{q} - \mathbf{r}(\mathbf{x})) \qquad (V.16)$$

By using the following matrix identities we obtain:

$$\Gamma^{-1} - \Gamma^{-1}\Sigma_k(\Gamma + \Sigma_k)^{-1} = \Gamma^{-1}(I - \Sigma_k(\Sigma_k + \Gamma)^{-1})$$
$$= \Gamma^{-1}((\Sigma_k + \Gamma) - \Sigma_k)(\Sigma_k + \Gamma)^{-1} = (\Sigma_k + \Gamma)^{-1} \qquad (V.17)$$

and similarly

$$\Sigma_k^{-1} - \Sigma_k^{-1}\Gamma(\Gamma + \Sigma_k)^{-1} = (\Sigma_k + \Gamma)^{-1} \qquad (V.18)$$

The solution for $d(\mathbf{x}, k)$ using (V.12), (V.16), (V.17) and (V.18) is:

$$d(\mathbf{x}, k) = (\mathbf{z} - \mathbf{q} - \mathbf{r}(\mathbf{x}) - \mathbf{c}[k])^T(\Gamma + \Sigma_k)^{-1}(\mathbf{z} - \mathbf{q} - \mathbf{r}(\mathbf{x}) - \mathbf{c}[k]) \qquad (V.19)$$

The relationship between the determinants of the covariance matrices is:

$$|\Gamma^{-1} + \Sigma_k^{-1}| = |\Gamma^{-1}(\Gamma + \Sigma_k)\Sigma_k^{-1}| = |\Gamma|^{-1}|\Gamma + \Sigma_k||\Sigma_k|^{-1} \qquad (V.20)$$

or alternatively

$$|\Gamma|^{1/2}|\Sigma_k|^{1/2} = |(\Gamma^{-1} + \Sigma_k^{-1})^{-1}|^{1/2}|\Gamma + \Sigma_k|^{1/2} \qquad (V.21)$$

so, by using (V.10), (V.15), (V.19) and (V.21), the product of Gaussians in (V.9) has the form:

$$p(\mathbf{z}|\mathbf{x}, \mathbf{n}, \mathbf{q}, k)\,p(\mathbf{x}|k)$$
$$= N_{\mathbf{x}}(\mathbf{b}(\mathbf{x}, k), (\Gamma^{-1} + \Sigma_k^{-1})^{-1})\,N_{\mathbf{z}}(\mathbf{q} + \mathbf{r}(\mathbf{x}) + \mathbf{c}[k], \Gamma + \Sigma_k) \qquad (V.22)$$

and this concludes our decomposition.

Integrating (V.22) with respect to $\mathbf{x}$ is not possible since $\mathbf{b}(\mathbf{x}, k)$ is a function of $\mathbf{x}$ through $\mathbf{r}(\mathbf{x})$. If we make the approximation that $\mathbf{r}(\mathbf{x}) = \mathbf{r}(\mathbf{c}_k) = \mathbf{r}[k]$ is constant for every mixture we can obtain easily

$$\int p(\mathbf{z} \mid \mathbf{x}, \mathbf{n}, \mathbf{q}, k)\, p(\mathbf{x} \mid k)\, d\mathbf{x} = N_{\mathbf{z}}(\mathbf{q} + \mathbf{r}[k] + \mathbf{c}[k], \Gamma + \Sigma_k) \qquad (V.23)$$

$$\int \mathbf{x}\, p(\mathbf{z} \mid \mathbf{x}, \mathbf{n}, \mathbf{q}, k)\, p(\mathbf{x} \mid k)\, d\mathbf{x} = \mathbf{b}[k] N_{\mathbf{z}}(\mathbf{q} + \mathbf{r}[k] + \mathbf{c}[k], \Gamma + \Sigma_k) \qquad (V.24)$$

with $\mathbf{b}[k]$ being given by

$$\mathbf{b}[k] = \Sigma_k (\Sigma_k + \Gamma)^{-1} (\mathbf{z} - \mathbf{q} - \mathbf{r}[k]) + \Gamma (\Sigma_k + \Gamma)^{-1}\, \mathbf{c}[k] \qquad (V.25)$$

Equations (V.23), (V.24) and (V.25) will be used in the derivation of the CDCN algorithm.

V.3. MMSE Estimate in CDCN

Using (V.23) and (V.24), Equation (V.8) can now be expressed as:

$$\hat{\mathbf{x}}_{MMSE} = \frac{\sum_{k=0}^{K-1} P[k]\, N_{\mathbf{z}}(\mathbf{q} + \mathbf{r}[k] + \mathbf{c}[k], \Gamma + \Sigma_k)\, \mathbf{b}[k]}{\sum_{k=0}^{K-1} P[k]\, N_{\mathbf{z}}(\mathbf{q} + \mathbf{r}[k] + \mathbf{c}[k], \Gamma + \Sigma_k)} \qquad (V.26)$$

or better as

$$\hat{\mathbf{x}}_{MMSE} = \sum_{k=0}^{K-1} f[k]\, \mathbf{b}[k] \qquad (V.27)$$

where $\mathbf{b}[k]$ are given by (V.25), and $f[k]$ are given by:

$$f[k] = \frac{P[k]\, N_{\mathbf{z}}(\mathbf{q} + \mathbf{r}[k] + \mathbf{c}[k], \Gamma + \Sigma_k)}{\sum_{l=0}^{K-1} P[l]\, N_{\mathbf{z}}(\mathbf{q} + \mathbf{r}[l] + \mathbf{c}[l], \Gamma + \Sigma_l)} \qquad (V.28)$$

To simplify (V.25) and (V.28) we will now make the assumptions described in Appendix IV, in which the spectral (cepstral) distance $d = \|\mathbf{z} - \mathbf{y}\|$ between the true PCD (Power Cepstral Density) $\mathbf{y}$ and the observed cepstrum $\mathbf{z}$ has a Gaussian distribution

$$f_{\mathbf{z}}(d) = \frac{1}{\sqrt{2\pi}\,\gamma} \exp\left(-\frac{d^2}{2\gamma^2}\right) \qquad (V.29)$$

In this case the cepstral distance takes the following value

$$d[k] = \| \mathbf{z} - \mathbf{q} - \mathbf{r}[k] - \mathbf{c}[k] \| \tag{V.30}$$

Furthermore, we will make the same assumption of Gaussian distribution of the cepstral distance for $p(\mathbf{x} \mid k)$, where all codewords share the same variance, except for the noise codeword. In addition, σ^2 will include the contribution of the codeword variance and the cepstral estimation variance. With this, Equation (V.28) can be expressed as

$$f[k] = \frac{\frac{P[k]}{\sigma} \exp(-\frac{d^2[k]}{2\sigma^2})}{\frac{P[0]}{\gamma} \exp(-\frac{d^2[0]}{2\gamma^2}) + \sum_{k=1}^{K-1} \frac{P[k]}{\sigma} \exp(-\frac{d^2[k]}{2\sigma^2})} \quad \text{for } k \neq 0 \tag{V.31}$$

with $d[k]$ given by (V.30) and

$$f[0] = \frac{\frac{P[0]}{\gamma} \exp(-\frac{d^2[0]}{2\gamma^2})}{\frac{P[0]}{\gamma} \exp(-\frac{d^2[0]}{2\gamma^2}) + \sum_{k=1}^{K-1} \frac{P[k]}{\sigma} \exp(-\frac{d^2[k]}{2\sigma^2})} \quad \text{for } k = 0 \tag{V.32}$$

with $d[0]$ given by

$$d[0] = \| \mathbf{z} - \mathbf{n} - \mathbf{s}[0] \| = \| \mathbf{z} - \mathbf{n} \| \tag{V.33}$$

in which we set the correction for the noise codeword s[0] to 0.

Finally we will assume that $\sigma^2 >> \gamma^2$, which was confirmed experimentally. With this, Equation (V.25) takes the form

$$\begin{aligned} \mathbf{b}[k] &= \mathbf{z} - \mathbf{q} - \mathbf{r}[k] \quad \text{for } k \neq 0 \\ \mathbf{b}[0] &= \mathbf{z} - \mathbf{n} \quad \text{for } k = 0 \end{aligned} \tag{V.34}$$

Equations (V.27), (V.34), (V.31), (V.32), (V.30) and (V.33) constitute the basis of the CDCN algorithm.

Appendix VI
Maximum Likelihood via the EM Algorithm

The EM (Estimate-Maximize) algorithm is a general method to solve maximum likelihood problems with incomplete data. It was developed by Laird *et al.* [45]. First we present the Jensen's inequality that is used in deriving the EM algorithm.

VI.1. Jensen's Inequality

Jensen's inequality deals with expectations of convex functions. In this subsection we will derive it for the specific case of the logarithm which is of interest for the EM algorithm.

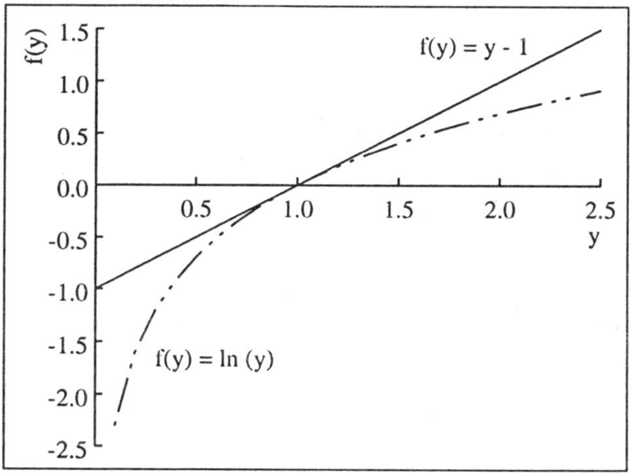

Figure VI-1: Functions $f(y) = y - 1$ and $f(y) = \ln(y)$.

It is clear from inspection of Figure VI-1 that

$$y - 1 \geq \ln y, \quad \forall y \tag{VI.1}$$

Let $f_{\mathbf{X}}(\mathbf{x})$ and $g_{\mathbf{X}}(\mathbf{x})$ be two legitimate pdfs so that

$$\int f_{\mathbf{X}}(\mathbf{x})\, d\mathbf{x} = \int g_{\mathbf{X}}(\mathbf{x})\, d\mathbf{x} = 1 \tag{VI.2}$$

or after some manipulation

$$0 = \int [f_{\mathbf{X}}(\mathbf{x}) - g_{\mathbf{X}}(\mathbf{x})]\, d\mathbf{x} = \int f_{\mathbf{X}}(\mathbf{x}) \left[\frac{g_{\mathbf{X}}(\mathbf{x})}{f_{\mathbf{X}}(\mathbf{x})} - 1 \right] d\mathbf{x} \qquad (VI.3)$$

for all vectors $\mathbf{x}$ whose $f_{\mathbf{X}}(\mathbf{x}) \neq 0$. For those vectors we have that

$$y = \frac{g_{\mathbf{X}}(\mathbf{x})}{f_{\mathbf{X}}(\mathbf{x})} \geq 0 \qquad (VI.4)$$

since both $g_{\mathbf{X}}(\mathbf{x}) \geq 0$ and $f_{\mathbf{X}}(\mathbf{x}) \geq 0$ for two legitimate pdfs. Substituting (VI.4) in (VI.1) we obtain

$$\frac{g_{\mathbf{X}}(\mathbf{x})}{f_{\mathbf{X}}(\mathbf{x})} - 1 \geq \ln \frac{g_{\mathbf{X}}(\mathbf{x})}{f_{\mathbf{X}}(\mathbf{x})} \qquad (VI.5)$$

Since $f_{\mathbf{X}}(\mathbf{x}) \geq 0$, multiplying both sides of (VI.5) by $f_{\mathbf{X}}(\mathbf{x})$ maintains the inequality sign, yielding:

$$f_{\mathbf{X}}(\mathbf{x}) \left[\frac{g_{\mathbf{X}}(\mathbf{x})}{f_{\mathbf{X}}(\mathbf{x})} - 1 \right] \geq f_{\mathbf{X}}(\mathbf{x}) \ln \frac{g_{\mathbf{X}}(\mathbf{x})}{f_{\mathbf{X}}(\mathbf{x})} \qquad (VI.6)$$

that after integration with respect to $\mathbf{x}$ combined with (VI.3) gives:

$$0 \geq \int f_{\mathbf{X}}(\mathbf{x}) \ln \frac{g_{\mathbf{X}}(\mathbf{x})}{f_{\mathbf{X}}(\mathbf{x})} d\mathbf{x} \qquad (VI.7)$$

or finally:

$$\int f_{\mathbf{X}}(\mathbf{x}) \ln f_{\mathbf{X}}(\mathbf{x})\, d\mathbf{x} \geq \int f_{\mathbf{X}}(\mathbf{x}) \ln g_{\mathbf{X}}(\mathbf{x})\, d\mathbf{x} \qquad (VI.8)$$

which is the version of Jensen's inequality that we will use here.

VI.2. The EM Algorithm

The derivation of the EM algorithm presented here is a modification of the one used by Feder and Weinstein [21]. Considerably more detail and proofs are in the paper by Laird *et al.* [45].

Let's define $\mathbf{Z}$ as the observed data, $\mathbf{X}$ as the unobserved data and θ a parameter vector. We can express the densities as

$$f_{\mathbf{XZ}}(\mathbf{x}, \mathbf{z}; \theta) = f_{\mathbf{X/Z=z}}(\mathbf{x}; \theta) f_{\mathbf{Z}}(\mathbf{z}; \theta) \qquad (VI.9)$$

where $f_{\mathbf{XZ}}(\mathbf{x}, \mathbf{z}; \theta)$ is the *joint pdf* of the complete data, $f_{\mathbf{X/Z=z}}(\mathbf{x}; \theta)$ is the conditional pdf of the unobserved data $\mathbf{X}$ given the observed data $\mathbf{z}$ and $f_{\mathbf{Z}}(\mathbf{z}; \theta)$ is the pdf of the observed data $\mathbf{z}$.

Now taking logarithms of (VI.9) we can obtain:

$$\ln f_{\mathbf{Z}}(\mathbf{z};\theta) = \ln f_{\mathbf{XZ}}(\mathbf{x},\mathbf{z};\theta) - \ln f_{\mathbf{X/Z=z}}(\mathbf{x};\theta) \qquad (VI.10)$$

Let's take conditional expectations given $\mathbf{Z}=\mathbf{z}$ and the parameter θ' on (VI.10):

$$\ln f_{\mathbf{Z}}(\mathbf{z};\theta) = E\{\ln f_{\mathbf{XZ}}(\mathbf{x},\mathbf{z};\theta) \mid \mathbf{Z}=\mathbf{z};\theta'\}$$

$$- E\{\ln f_{\mathbf{X/Z=z}}(\mathbf{x};\theta) \mid \mathbf{Z}=\mathbf{z};\theta'\} \qquad (VI.11)$$

and define for convenience the following quantities:

$$L(\theta) = \ln f_{\mathbf{Z}}(\mathbf{z};\theta) \qquad (VI.12)$$

$$U(\theta,\theta') = E\{\ln f_{\mathbf{XZ}}(\mathbf{x},\mathbf{z};\theta) \mid \mathbf{Z}=\mathbf{z};\theta'\} \qquad (VI.13)$$

$$V(\theta,\theta') = E\{\ln f_{\mathbf{X/Z=z}}(\mathbf{x};\theta) \mid \mathbf{Z}=\mathbf{z};\theta'\} \qquad (VI.14)$$

so that (VI.11) can be written in a compact form as:

$$L(\theta) = U(\theta,\theta') - V(\theta,\theta') \qquad (VI.15)$$

Applying the Jensen's inequality of (VI.8) to (VI.14) it is clear that

$$V(\theta,\theta') \le V(\theta',\theta') \qquad (VI.16)$$

Hence if we find a value of the parameter vector θ that makes

$$U(\theta,\theta') > U(\theta',\theta') \qquad (VI.17)$$

then a combination of (VI.15), (VI.16) and (VI.17) lead to

$$L(\theta) > L(\theta') \qquad (VI.18)$$

With these relationships we have converted the problem of maximizing the log-likelihood $L(\theta)$ to the one of maximizing the function $U(\theta,\theta')$ defined by (VI.13). The EM algorithm starts with an initial estimate for $\hat{\theta}^{(0)}$. Let us denote $\hat{\theta}^{(n)}$ the current estimate of θ after n iterations. The iteration can be described in two steps:

1. *Estimate* $U(\theta,\hat{\theta}^{(n)})$ by taking the expectation over the unobserved data.
2. *Maximize* $U(\theta,\hat{\theta}^{(n)})$ to obtain $\hat{\theta}^{(n+1)}$

It is proved in [45] (Laird *et al.* that if $U(\theta,\theta')$ is continuous in both θ and θ', the algorithm converges to stationary point of the log-likelihood function where the maximization ensures that each iteration increases the likelihood. As in any "hill-climbing" method, the algorithm may converge to a local maximum rather than to the global maximum.

Appendix VII
ML Estimation of Noise and Spectral Tilt

In this appendix we apply the EM algorithm described in Appendix VI to the problem of maximum likelihood estimation of the noise $\mathbf{n}$ and spectral equalization $\mathbf{q}$ given the set of observations $\{\mathbf{z}_i;\ i=0,1,\ldots,N-1\}$ in the context of the CDCN algorithm. The unobserved data will be the mixture k and the clean vectors $\{\mathbf{x}_i\}$. With these definitions, the function $U(\mathbf{n},\mathbf{q},\mathbf{n}',\mathbf{q}')$ defined in (VI.13) has the form:

$$U = \sum_{i=0}^{N-1}\sum_{k=0}^{K-1}\int p(\mathbf{x}_i, k \mid \mathbf{z}_i, \mathbf{n}', \mathbf{q}') \ln p(\mathbf{x}_i, k, \mathbf{z}_i \mid \mathbf{n}, \mathbf{q})\, d\mathbf{x}_i \tag{VII.1}$$

where the first term in the integral has the form

$$p(\mathbf{x}_i, k \mid \mathbf{z}_i, \mathbf{n}', \mathbf{q}') = \frac{p(\mathbf{z}_i, \mathbf{x}_i, k \mid \mathbf{n}', \mathbf{q}')}{p(\mathbf{z}_i \mid \mathbf{n}', \mathbf{q}')} \tag{VII.2}$$

whose numerator can be expressed as

$$p(\mathbf{z}_i, \mathbf{x}_i, k \mid \mathbf{n}', \mathbf{q}') = p(\mathbf{z}_i \mid \mathbf{x}_i, k, \mathbf{n}', \mathbf{q}')\, p(\mathbf{x}_i, k \mid \mathbf{n}', \mathbf{q}')$$

$$= p(\mathbf{z}_i \mid \mathbf{x}_i, k, \mathbf{n}', \mathbf{q}')\, p(\mathbf{x}_i \mid k)\, P[k] \tag{VII.3}$$

Integration on $\mathbf{x}_i$ and summation on k in Equation (VII.3) yields the denominator in (VII.2):

$$p(\mathbf{z}_i \mid \mathbf{n}', \mathbf{q}') = \sum_{k=0}^{K-1} P[k] \int p(\mathbf{z}_i \mid \mathbf{x}_i, k, \mathbf{n}', \mathbf{q}')\, p(\mathbf{x}_i \mid k)\, d\mathbf{x}_i \tag{VII.4}$$

So that (VII.1) takes on the new form

$$U = \sum_{i=0}^{N-1} \frac{\sum_{k=0}^{K-1} P[k] \int p(\mathbf{z}_i \mid \mathbf{x}_i, k, \mathbf{n}', \mathbf{q}')\, p(\mathbf{x} \mid k) \ln p(\mathbf{z}_i, \mathbf{x}_i, k \mid \mathbf{n}, \mathbf{q})\, d\mathbf{x}_i}{\sum_{k=0}^{K-1} P[k] \int p(\mathbf{z}_i \mid \mathbf{x}_i, k, \mathbf{n}', \mathbf{q}')\, p(\mathbf{x}_i \mid k)\, d\mathbf{x}_i} \tag{VII.5}$$

Also, the term with the logarithm in (VII.5) can be expressed as

$$p(\mathbf{z}_i, \mathbf{x}_i, k \mid \mathbf{n}, \mathbf{q}) = p(\mathbf{z}_i \mid \mathbf{x}_i, k, \mathbf{n}, \mathbf{q})\, p(\mathbf{x}_i \mid k)\, P[k] \tag{VII.6}$$

where the term depending on $\mathbf{n}$ and $\mathbf{q}$ can be expressed as

$$\ln p(\mathbf{z}_i | \mathbf{x}_i, k, \mathbf{n}, \mathbf{q})$$
$$= \alpha' - \frac{1}{2}(\mathbf{z}_i - \mathbf{q} - \mathbf{x}_i - \mathbf{r}[k])^T \Gamma^{-1} (\mathbf{z}_i - \mathbf{q} - \mathbf{x}_i - \mathbf{r}[k])$$
$$= \alpha' - \frac{1}{2}(\mathbf{z}_i - \mathbf{x}_i)^T \Gamma^{-1} (\mathbf{z}_i - \mathbf{x}_i) - \frac{1}{2}(\mathbf{q} + \mathbf{r}[k])^T \Gamma^{-1} (\mathbf{q} + \mathbf{r}[k])$$
$$+ (\mathbf{z}_i - \mathbf{x}_i)^T \Gamma^{-1} (\mathbf{q} + \mathbf{r}[k]) \qquad (VII.7)$$

where we have used the approximation that the correction vectors are constant within a cluster:

$$\mathbf{r}(\mathbf{x}_i) = \mathbf{r}[k] \quad \forall i \qquad (VII.8)$$

Equation (VII.7) can be expanded as

$$\ln p(\mathbf{z}_i | \mathbf{x}_i, k, \mathbf{n}, \mathbf{q}) = \alpha' - \frac{1}{2}(\mathbf{z}_i - \mathbf{x}_i)^T \Gamma^{-1} (\mathbf{z}_i - \mathbf{x}_i)$$
$$- \frac{1}{2}(\mathbf{q} + \mathbf{r}[k])^T \Gamma^{-1} [\mathbf{q} + \mathbf{r}[k] - 2(\mathbf{z}_i - \mathbf{x}_i)] \qquad (VII.9)$$

or in terms of the **s** correction vectors as

$$\ln p(\mathbf{z}_i | \mathbf{x}_i, k, \mathbf{n}, \mathbf{q}) = \alpha' - \frac{1}{2}(\mathbf{z}_i - \mathbf{n} - \mathbf{s}[k])^T \Gamma^{-1} (\mathbf{z}_i - \mathbf{n} - \mathbf{s}[k]) \qquad (VII.10)$$

that for the noise codeword takes on the value

$$\ln p(\mathbf{z}_i | \mathbf{x}_i, 0, \mathbf{n}, \mathbf{q}) = \alpha' - \frac{1}{2}(\mathbf{z}_i - \mathbf{n})^T \Gamma^{-1} (\mathbf{z}_i - \mathbf{n}) \qquad (VII.11)$$

since $\mathbf{s}[0] \approx 0$.

Using (V.23), (V.24) and (V.28), Equation (VII.5) can be expressed as

$$U = \alpha'' - \frac{1}{2}\sum_{i=0}^{N-1} f_i[0]\,(\mathbf{z}_i - \mathbf{n})^T \Gamma^{-1} (\mathbf{z}_i - \mathbf{n})$$
$$- \frac{1}{2}\sum_{i=0}^{N-1} \sum_{k=1}^{K-1} f_i[k]\,(\mathbf{q} + \mathbf{r}[k])^T \Gamma^{-1} [\mathbf{q} + \mathbf{r}[k] - 2(\mathbf{z}_i - \mathbf{b}'_i[k])] \qquad (VII.12)$$

where α'' includes terms that do not depend on either $\mathbf{q}$ nor $\mathbf{n}$. From Equation (V.25) $\mathbf{b}'_i[k]$ is given by

$$\mathbf{b}'_i[k] = \Sigma_k (\Sigma_k + \Gamma)^{-1} (\mathbf{z}_i - \mathbf{q}' - \mathbf{r}'[k]) + \Gamma (\Sigma_k + \Gamma)^{-1} \mathbf{c}[k] \qquad (VII.13)$$

Now, taking the derivative with respect to $\mathbf{n}$ in (VII.12)

$$\frac{\delta U}{\delta \mathbf{n}} = \sum_{i=0}^{N-1} f_i[0]\,\Gamma^{-1} (\mathbf{z}_i - \mathbf{n}) = 0 \qquad (VII.14)$$

Solving for **n** we obtain the ML estimate

$$\hat{\mathbf{n}} = \frac{\sum_{i=0}^{N-1} f_i[0]\, \mathbf{z}_i}{\sum_{i=0}^{N-1} f_i[0]} \qquad (VII.15)$$

Similarly, taking the derivative with respect to **q** in (VII.12)

$$\frac{\delta U}{\delta \mathbf{q}} = \sum_{i=0}^{N-1} \sum_{k=1}^{K-1} f_i[k]\, (I + \frac{\delta \mathbf{r}[k]}{\delta \mathbf{q}}) \Gamma^{-1} (\mathbf{z}_i - \mathbf{q} - \mathbf{r}[k] - \mathbf{b}_i'[k]) = 0 \qquad (VII.16)$$

We cannot obtain a closed-form expression for Equation (VII.16) in this form. Fortunately we can see that the term $\delta \mathbf{r}[k] / \delta \mathbf{q}$ is negligible compared with I for speech codewords[35].

$$(I + \frac{\delta \mathbf{r}[k]}{\delta \mathbf{q}}) \Gamma^{-1} (\mathbf{z}_i - \mathbf{q} - \mathbf{r}[k] - \mathbf{b}_i'[k]) \approx \Gamma^{-1} (\mathbf{z}_i - \mathbf{q} - \mathbf{r}[k] - \mathbf{b}_i'[k]) \qquad (VII.17)$$

We will approximate $\mathbf{r}[k]$ by $\mathbf{r}'[k]$, which entails to use the previous correction vectors:

$$\begin{aligned} &\mathbf{z}_i - \mathbf{q} - \mathbf{r}[k] - \mathbf{b}_i'[k] \\ &= \mathbf{z}_i - \mathbf{q} - \mathbf{r}'[k] - \mathbf{b}_i'[k] = (\mathbf{q}' - \mathbf{q}) + (\mathbf{z}_i - \mathbf{q}' - \mathbf{r}'[k] - \mathbf{b}_i'[k]) \\ &= (\mathbf{q}' - \mathbf{q}) + \Gamma(\Sigma_k + \Gamma)^{-1} (\mathbf{z}_i - \mathbf{q}' - \mathbf{r}'[k] - \mathbf{c}[k]) \\ &= -\mathbf{q} + \Sigma_k (\Sigma_k + \Gamma)^{-1} \mathbf{q}' + \Gamma(\Sigma_k + \Gamma)^{-1} (\mathbf{z}_i - \mathbf{r}'[k] - \mathbf{c}[k]) \end{aligned} \qquad (VII.18)$$

where we have made use of (VII.13), (V.13) and (V.14).

Solving for **q** in (VII.16) with the use of (VII.17) and (VII.18) we obtain

$$\hat{\mathbf{q}} = \Sigma(\Sigma + \Gamma)^{-1} \mathbf{q}' + \Gamma(\Sigma + \Gamma)^{-1} \frac{\sum_{i=0}^{N-1} \sum_{k=1}^{K-1} f_i[k]\, (\mathbf{z}_i - \mathbf{r}'[k] - \mathbf{c}[k])}{\sum_{i=0}^{N-1} \sum_{k=1}^{K-1} f_i[k]} \qquad (VII.19)$$

where we have assumed that $\Sigma_k = \Sigma \quad \forall\, k$.

---

[35]Note that the function $f(x) = \ln(1 + \exp(x))$ is much smaller than 1 for $x < -2$. In addition, its derivative, $f'(x) = 1/(1 + \exp(-x))$, is also much smaller than 1 for $x < -2$. This function is used in the correction vectors, where for speech codewords this approximation is highly valid.

Equation (VII.19) has a contribution of the former $\mathbf{q}'$ and a new component. If in the EM algorithm, as described in Appendix VI, the new parameter vector is $\theta = (\mathbf{q}, \mathbf{n}, \mathbf{r}')$ instead of $\theta = (\mathbf{q}, \mathbf{n}, \mathbf{r})$, we could reestimate $\mathbf{q}$ in (VII.19) as many times as we wanted without recomputing $\mathbf{r}[k]$. By doing so, we can substitute $\mathbf{q}'$ by $\hat{\mathbf{q}}$ in (VII.19) which would give us after using (V.13) and (V.14) the following equation:

$$\hat{\mathbf{q}} = \frac{\sum_{i=0}^{N-1} \sum_{k=1}^{K-1} f_i[k] \, (\mathbf{z}_i - \mathbf{c}[k] - \mathbf{r}'[k])}{\sum_{i=0}^{N-1} \sum_{k=1}^{K-1} f_i[k]} \qquad (VII.20)$$

We have found in practice, that the (VII.20) converges faster than (VII.19).

Equations (VII.15) and (VII.20) give improved estimates of $\mathbf{n}$ and $\mathbf{q}$, in the sense of a higher likelihood.

Appendix VIII
Vocabulary and Pronunciation Dictionary

A	EY
AND	EH N DD
APOSTROPHE	AX P AA S T R OW F IY
APRIL	EY P R L
AREA	EH R IY AX
AUGUST	AO G AX S TD
B	B IY
C	S IY
CODE	K OW DD
D	D IY
DECEMBER	D IX S EH M B ER
E	IY
EIGHT	EY TD
EIGHTEEN	EY T IY N
EIGHTEENTH	EY T IY N TH
EIGHTH	EY TH
EIGHTY	EY DX IY
ELEVEN	AX L EH V IH N
ELEVENTH	AX L EH V IH N TH
ENTER	EH N T ER
ERASE	IX R EY S
F	EH F
FEBRUARY	F EH B Y UW EH R IY
FIFTEEN	F IH F T IY N
FIFTEENTH	F IH F T IY N TH
FIFTH	F IH F TH
FIFTY	F IH F T IY
FIRST	F ER S TD
FIVE	F AY V
FORTY	F AO R DX IY
FOUR	F AO R
FOURTEEN	F AO R T IY N
FOURTH	F AO R TH
G	JH IY
GO	G OW
H	EY CH
HALF	HH AE F
HALL	HH AA L
HELP	HH EH L PD
HUNDRED	HH AH N D R AX DD
I	AY
J	JH EY

JANUARY	JH AE N Y UW EH R IY
JULY	JH AX L AY
JUNE	JH UW N
K	K EY
L	EH L
LANE	L EY N
M	EH M
MARCH	M AA R CH
MAY	M EY
MEMORY	M EH M R IY
N	EH N
NINE	N AY N
NINETEEN	N AY N T IY N
NINETY	N AY N DX IY
NINTH	N AY N TH
NO	N OW
O	OW
OCTOBER	AA KD T OW B ER
OF	AX V
OH	OW
ONE	W AH N
P	P IY
Q	K Y UW
R	AA R
REPEAT	R IX P IY TD
RUBOUT	R AH B AW TD
S	EH S
SECOND	S EH K AX N DD
SEPTEMBER	S EH PD T EH M B ER
SEVEN	S EH V AX N
SEVENTEEN	S EH V AX N T IY N
SEVENTH	S EH V AX N TH
SEVENTY	S EH V AX N DX IY
SIL	SIL
SIX	S IH K S
SIXTEEN	S IH K S T IY N
SIXTEENTH	S IH K S T IY N TH
SIXTH	S IH K S TH
SIXTY	S IH K S T IY
START	S T AA R TD
STOP	S T AA PD
T	T IY
TEN	T EH N
THIRD	TH ER DD
THIRTIETH	TH ER DX IY IX TH
THIRTY	TH ER DX IY

THOUSAND	TH AW Z AX N DD
THREE	TH R IY
TWELFTH	T W EH L F TH
TWELVE	T W EH L V
TWENTIETH	T W EH N IY IX TH
TWENTY	T W EH N IY
TWO	T UW
U	Y UW
V	V IY
W	D AH B AH L Y UW
WEAN	W IY N
X	EH K S
Y	W AY
YES	Y EH S
Z	Z IY
ZERO	Z IY R OW

References

[1] A. Acero and R. M. Stern.
Environmental Robustness in Automatic Speech Recognition.
In *Proc. IEEE Int. Conf. Acoustics, Speech and Signal Processing, Albuquerque, NM*, pages 849-852. April, 1990.

[2] A. Acero and R. M. Stern.
Robust Speech Recognition by Normalization of the Acoustic Space.
In *Proc. IEEE Int. Conf. Acoustics, Speech and Signal Processing, Toronto, Canada*, pages 893-896. May, 1991.

[3] B. S. Atal.
Effectiveness of Linear Prediction Characteristics of the Speech Wave for Automatic Speaker Identification and Verification.
Journal of the Acoustical Society of America 55:1304-1312, Jun., 1974.

[4] M. Berouti, R. Schwartz and J. Makhoul.
Enhancement of Speech Corrupted by Acoustic Noise.
In J. S. Lim (editor), *Signal Processing*. Volume 1: *Speech Enhancement*, pages 69-73. Prentice Hall, Englewood Cliffs, NJ, 1983.

[5] B. Bogert, M. Healy and J. Tukey.
The Quefrency Alanysis of Time Series for Echoes.
In *Proc. Symp. on Time Series Analysis*, pages 209-243. J. Wiley, 1963.

[6] S. F. Boll.
Suppression of Acoustic Noise in Speech Using Spectral Subtraction.
IEEE Trans. Acoustics, Speech and Signal Processing 27(2):113-120, April, 1979.

[7] S. Boll, J. Porter and L. G. Bahler.
Robust Syntax Free Speech Recognition.
In *Proc. IEEE Int. Conf. Acoustics, Speech and Signal Processing, New York, NY*, pages 179-182. 1988.

[8] Y. Chen.
Cepstral Domain Talker Stress Compensation for Robust Speech Recognition.
IEEE Trans. Acoustics, Speech and Signal Processing ASSP-36:433-439, Apr., 1988.

[9] Chow, Y.L., Dunham, M.O., Kimball, O.A., Krasner, M.A., Kubala, G.F., Makhoul, J., Roucos, S., Schwartz, R.M.
BYBLOS: The BBN Continuous Speech Recognition System.
In *Proc. IEEE Int. Conf. Acoustics, Speech and Signal Processing, Dallas, TX*, pages 89-92. April, 1987.

[10] J. R. Cohen.
Application of an Auditory Model to Speech Recognition.
Journal of the Acoustical Society of America 85:2623-2629, Jun., 1989.

[11] D. Van Compernolle.
Spectral Estimation Using a Log-Distance Error Criterion Applied to Speech Recognition.
In *Proc. IEEE Int. Conf. Acoustics, Speech and Signal Processing, Glasgow, UK*, pages 258-261. May, 1989.

[12] D. Van Compernolle.
Noise Adaptation in a Hidden Markov Model Speech Recognition System.
Computer Speech and Language , 1989.

[13] D. Van Compernolle.
Switching Adaptive Filters for Enhancing Noisy and Reverberant Speech from Microphone Array Recordings.
In *Proc. IEEE Int. Conf. Acoustics, Speech and Signal Processing, Albuquerque, NM*, pages 833-836. April, 1990.

[14] S. B. Davis and P. Mermelstein.
Comparison of Parametric representations for Monosyllabic Word Recognition in Continuously Spoken Sentences.
IEEE Trans. Acoustics, Speech and Signal Processing ASSP-28(4):375-366, Aug., 1980.

[15] Y. Ephraim and D. Malah.
Speech Enhancement Using a Minimum Mean-Square Error Short-Time Spectral Amplitude Estimator.
IEEE Trans. Acoustics, Speech and Signal Processing ASSP-32(6):1109-1121, Dec., 1984.

[16] Y. Ephraim and D. Malah.
Speech Enhancement Using a Minimum Mean-Square Error Log-Spectral Amplitude Estimator.
IEEE Trans. Acoustics, Speech and Signal Processing ASSP-33(2):443-445, Apr., 1985.

[17] Y. Ephraim, D. Malah and B. H. Huang.
Speech Enhancement Based upon Hidden Markov Modeling.
In *Proc. IEEE Int. Conf. Acoustics, Speech and Signal Processing, Glasgow, UK*, pages 353-356. May, 1989.

[18] Y. Ephraim.
A Minimum Mean Square Error Approach for Speech Enhancement.
In *Proc. IEEE Int. Conf. Acoustics, Speech and Signal Processing, Albuquerque, NM*, pages 829-832. April, 1990.

[19] A. Erell and M. Weintraub.
Spectral Estimation for Noise Robust Speech Recognition.
In *Proc. Speech and Natural Language Workshop, Cape Cod, MA*. Morgan Kaufmann, Oct., 1989.

[20] A. Erell and M. Weintraub.
Estimation Using Log-Spectral-Distance Criterion for Noise-Robust Speech Recognition.
In *Proc. IEEE Int. Conf. Acoustics, Speech and Signal Processing, Albuquerque, NM*, pages 853-856. April, 1990.

[21] M. Feder and E. Weinstein.
Parameter Estimation of Superimposed Signals Using the EM Algorithm.
IEEE Trans. Acoustics, Speech and Signal Processing ASSP-36:477-489, Apr., 1988.

[22] M. Feder, A. V. Oppenheim and E. Weinstein.
Maximum Likelihood Noise Cancellation Using the EM Algorithm.
IEEE Trans. Acoustics, Speech and Signal Processing ASSP-37(2):204-216, Feb., 1989.

[23] J. L. Flanagan, J. D. Johnston, R. Zahn and G.W. Elko.
Computer-steered Microphone Arrays for Sound Transduction in Large Rooms.
Journal of the Acoustical Society of America 78:1508-1518, Nov., 1985.

[24] S. Furui.
Cepstral Analysis Technique for Automatic Speaker Verification.
IEEE Trans. Acoustics, Speech and Signal Processing ASSP-29(2):254-272, Apr., 1981.

[25] S. Furui.
Unsupervised Speaker Adaptation Method Based on Hierarchical Spectral Clustering.
In *Proc. IEEE Int. Conf. Acoustics, Speech and Signal Processing, Glasgow, UK*, pages 286-289. May, 1989.

[26] O. Ghitza.
Auditory Nerve Representation Criteria for Speech Analysis/Synthesis.
IEEE Trans. Acoustics, Speech and Signal Processing ASSP-35:736-740, Jun., 1987.

[27] O. Ghitza.
Auditory Neural Feedback as a Basis for Speech Processing.
In *Proc. IEEE Int. Conf. Acoustics, Speech and Signal Processing, New York, NY*, pages 91-94. 1988.

[28] L. Gillick.
Some Statistical Issues in the Comparison of Speech Recognition Algorithms.
In *Proc. IEEE Int. Conf. Acoustics, Speech and Signal Processing, Glasgow, UK*, pages 532-535. May, 1989.

[29] H. Gish, Y. Chow and J. R. Rohlicek.
Probabilistic Vector Mapping of Noisy Speech Parameters for HMM Word Spotting.
In *Proc. IEEE Int. Conf. Acoustics, Speech and Signal Processing, Albuquerque, NM*, pages 117-120. April, 1990.

[30] V. Goncharoff and S. Chandran.
Adaptive Speech Modification by Spectral Warping.
In *Proc. IEEE Int. Conf. Acoustics, Speech and Signal Processing, New York, NY*, pages 343-346. 1988.

[31] R. Gray, A. Buzo, A. Gray and Y. Matusyama.
Distance Measures for Speech Processing.
IEEE Trans. Acoustics, Speech and Signal Processing ASSP-24:380-391, Oct., 1976.

[32] Gray, R.M.
Vector Quantization.
IEEE ASSP Magazine 1(2):4-29, April, 1984.

[33] J. H. Hansen and M. A. Clements.
Constrained Iterative Speech Enhancement with Application to Automatic Speech Recognition.
In *Proc. IEEE Int. Conf. Acoustics, Speech and Signal Processing, New York, NY*, pages 561-564. 1988.

[34] J. H. Hansen.
Stress Compensation and Noise Reduction Algorithms for Robust Speech Recognition.
In *Proc. IEEE Int. Conf. Acoustics, Speech and Signal Processing, Glasgow, UK*, pages 266-269. May, 1989.

[35] H. Hermansky.
Optimization of Perceptually-Based ASR Front-End.
In *Proc. IEEE Int. Conf. Acoustics, Speech and Signal Processing, New York, NY*, pages 219-222. 1988.

[36] H. W. Hon and K. F. Lee.
On Vocabulary-Independent Speech Modeling.
In *Proc. IEEE Int. Conf. Acoustics, Speech and Signal Processing, Albuquerque, NM*, pages 725-728. April, 1990.

[37] X. D. Huang, K. F. Lee and H. W. Hon.
On Semi-Continuous Hidden Markov Modeling.
In *Proc. IEEE Int. Conf. Acoustics, Speech and Signal Processing, Albuquerque, NM*, pages 689-692. April, 1990.

[38] M. J. Hunt and C. Lefebvre.
Speaker Dependent and Independent Speech Recognition Experiments with an Auditory Model.
In *Proc. IEEE Int. Conf. Acoustics, Speech and Signal Processing, New York, NY*, pages 215-218. 1988.

[39] M. J. Hunt and C. Lefebvre.
A comparison of Several Acoustic Representations for Speech Recognition with Degraded and Undegraded Speech.
In *Proc. IEEE Int. Conf. Acoustics, Speech and Signal Processing, Glasgow, UK*, pages 262-265. May, 1989.

[40] F. Itakura and T. Umezaki.
Distance Measure for Speech Recognition Based on the Smoothed Group Delay Spectrum.
In *Proc. IEEE Int. Conf. Acoustics, Speech and Signal Processing, Dallas, TX*, pages 1257-1260. 1987.

[41] N. S. Jayant and P. Noll.
Digital Coding of Waveforms.
Prentice Hall, 1984.

[42] B. H. Juang, L. R. Rabiner and J. G. Wilpon.
On the Use of Bandpass Liftering in Speech Recognition.
IEEE Trans. Acoustics, Speech and Signal Processing ASSP-35:947-954, Jul., 1987.

[43] J. C. Junqua and H. Wakita.
A Comparative Study of Cepstral Lifters and Distance Measures for All-Pole Models of Speech in Noise.
In *Proc. IEEE Int. Conf. Acoustics, Speech and Signal Processing, Glasgow, UK*, pages 476-479. May, 1989.

[44] S. M. Kay.
Modern Spectral Estimation.
Prentice Hall, 1988.

[45] N.M. Laird, A.P. Dempster and D.B. Rubin.
Maximum Likelihood from Incomplete Data via the EM algorithm.
Ann. Roy. Stat. Soc. :1-38, Dec, 1987.

[46] Lee, K.F.
Automatic Speech Recognition: The Development of the SPHINX System.
Kluwer Academic Publishers, Boston, 1989.

[47] K. F. Lee and H. W. Hon.
Large-Vocabulary Speaker-Independent Continuous Speech Recognition Using HMM.
In *Proc. IEEE Int. Conf. Acoustics, Speech and Signal Processing, New York, NY*, pages 123-126. 1988.

[48] K. F. Lee, H.W Hon, M.Y Hwang, S. Mahajan and R. Reddy.
The SPHINX Speech Recognition System.
In *Proc. IEEE Int. Conf. Acoustics, Speech and Signal Processing, Glasgow, UK*, pages 445-448. May, 1989.

[49] K. P. Li and J. Porter.
Normalizations and Selection of Speech Segments for Speaker Recognition Scoring.
In *Proc. IEEE Int. Conf. Acoustics, Speech and Signal Processing, New York, NY*, pages 595-598. 1988.

[50] J. S. Lim and A. V. Oppenheim.
All-Pole Modeling of Degraded Speech.
IEEE Trans. Acoustics, Speech and Signal Processing ASSP-26:197-210, Jun., 1978.

[51] J. S. Lim.
Speech Enhancement.
Prentice-Hall, Englewood Cliffs, NJ, 1983.

[52] Linde, Y., Buzo, A., Gray, R.M.
An Algorithm for Vector Quantizer Design.
IEEE Transactions on Communication COM-28(1):84-95, January, 1980.

[53] R. P. Lippmann, E. A. Martin and D.B. Paul.
Multi-Style Training for Robust Isolated-Word Speech Recognition.
In *Proc. IEEE Int. Conf. Acoustics, Speech and Signal Processing, Dallas, TX*, pages 705-708. April, 1987.

[54] F. H. Liu, A. Acero and R. M. Stern.
Efficient Joint Compensation of Speech for the Effects of Additive Noise and Linear Filtering.
In *Proc. IEEE Int. Conf. Acoustics, Speech and Signal Processing, San Francisco, CA*, pages 257-260. Mar, 1992.

[55] D. Mansour and B. H. Juang.
A Family of Distortion Measures Based Upon Projection Operation for Robust Speech Recognition.
In *Proc. IEEE Int. Conf. Acoustics, Speech and Signal Processing, New York, NY*, pages 36-39. 1988.

[56] K. S. Min, D. Chien, S. Li and C. Jones.
Automated Two-Speaker Separation System.
In *Proc. IEEE Int. Conf. Acoustics, Speech and Signal Processing, New York, NY*, pages 537-540. 1988.

[57] S. Morii.
Spectral Subtraction in the Sphinx System.
unpublished.
1988

[58] H. Murveit and M. Weintraub.
1000-Word Speaker-Independent Continuous-Speech Recognition Using Hidden Markov Models.
In *Proc. IEEE Int. Conf. Acoustics, Speech and Signal Processing, New York, NY*, pages 115-118. 1988.

[59] A. J. Nadas, D. Nahamoo and M. A. Picheny.
Speech Recognition Using Noise-Adaptive Prototypes.
In *Proc. IEEE Int. Conf. Acoustics, Speech and Signal Processing, New York, NY*, pages 517-520. 1988.

[60] A. J. Nadas, D. Nahamoo and M. A. Picheny.
Adaptive labeling: Normalization of Speech by Adaptive Transformations Based on Vector Quantization.
In *Proc. IEEE Int. Conf. Acoustics, Speech and Signal Processing, New York, NY*, pages 521-524. 1988.

[61] J. A. Naylor and S. F. Boll.
Techniques for Suppression of an Interfering Talker in Co-Channel Speech.
In *Proc. IEEE Int. Conf. Acoustics, Speech and Signal Processing, Dallas, TX*, pages 205-208. 1987.

[62] Nocerino, F. K. Soong, L. R. Rabiner and D. H. Klatt.
Comparative Study of Several Distortion Measures for Speech Recognition.
In *Proc. IEEE Int. Conf. Acoustics, Speech and Signal Processing, Atlanta, GA*, pages 25-28. Apr., 1985.

[63] H. Noda.
Frequency-Warped Spectral Distance Measures for Speaker Verification in Noise.
In *Proc. IEEE Int. Conf. Acoustics, Speech and Signal Processing, New York, NY*, pages 576-579. 1988.

[64] A. V. Oppenheim and D. H. Johnson.
Discrete Representation of Signals.
Proc. of the IEEE (33):681-691, 1972.

[65] D. S. Pallett.
Benchmark Tests for DARPA Resource Management Performance Evaluations.
In *Proc. IEEE Int. Conf. Acoustics, Speech and Signal Processing, Glasgow, UK*, pages 536-539. May, 1989.

[66] D. S. Pallett, W. M. Fisher and J. G. Fiscus.
Tools for the Analysis of Benchmark Speech Recognition Tests.
In *Proc. IEEE Int. Conf. Acoustics, Speech and Signal Processing, Albuquerque, NM*, pages 97-100. April, 1990.

[67] A. Papoulis.
Probability, Random Variables and Stochastic Processes.
Mc Graw Hill, 1984.

[68] D. B. Paul.
A Speaker-Stress Resistant Isolated Word Recognizer.
In *Proc. IEEE Int. Conf. Acoustics, Speech and Signal Processing, Dallas, TX*, pages 713-716. 1987.

[69] J. Picone.
Continuous Speech Recognition Using Hidden Markov Models.
IEEE ASSP Magazine 7(3):26-41, July, 1990.

[70] J. E. Porter and S. F. Boll.
Optimal Estimators for Spectral Restoration of Noisy Speech.
In *Proc. IEEE Int. Conf. Acoustics, Speech and Signal Processing, San Diego, CA*, pages 18A.2.1. May, 1984.

[71] L. R. Rabiner and R. W. Schafer.
Digital Processing of Speech Signals.
Prentice Hall, 1978.

[72] Rabiner, L.R., Juang, B.H.
An Introduction to Hidden Markov Models.
IEEE ASSP Magazine 3(1):4-16, January, 1986.

[73] H. Sakoe and S. Chiba.
Dynamic Programming Algorithm Optimization for Spoken Word Recognition.
IEEE Trans. Acoustics, Speech and Signal Processing ASSP-26:43-49, 1978.

[74] S. Seneff.
A Joint Synchrony/Mean-Rate Model of Auditory Speech Processing.
Journal of Phonetics 16:55-76, Jan., 1988.

[75] Shikano, K.
Evaluation of LPC Spectral Matching Measures for Phonetic Unit Recognition.
Technical Report, Computer Science Department, Carnegie Mellon University, May, 1986.

[76] H. Silverman.
An Algorithm for Determining Talker Location using a Linear Microphone Array and Optimal Hyperbolic Fit.
In *Proc. Speech and Natural Language Workshop, Hidden Valley, PA.* Morgan Kaufmann, Jun., 1990.

[77] F. K. Soong and M. M. Sondhi.
A Frequency-Weighted Itakura Spectral Distortion Measure and its Application to Speech Recognition in Noise.
In *Proc. IEEE Int. Conf. Acoustics, Speech and Signal Processing, Dallas, TX*, pages 625-628. 1987.

[78] R. Stern and A. Acero.
Acoustical Pre-processing for Robust Speech Recognition.
In *Proc. Speech and Natural Language Workshop, Cape Cod, MA*, pages 311-318. Morgan Kaufmann, Oct., 1989.

[79] T. G. Stockham, T. M. Cannon and R. B. Ingebretsen.
Blind Deconvolution Through Digital Signal Processing.
Proc. of the IEEE 63(4):678-692, Apr., 1975.

[80] S. Tamura and A. Waibel.
Noise Reduction Using Connectionist Models.
In *Proc. IEEE Int. Conf. Acoustics, Speech and Signal Processing, New York, NY*, pages 553-556. 1988.

[81] S. Tamura and M. Nakamura.
Improvements to the Noise Reduction Neural Network.
In *Proc. IEEE Int. Conf. Acoustics, Speech and Signal Processing, Albuquerque, NM*, pages 825-828. April, 1990.

[82] Y. Tokhura.
A Weighted Cepstral Distance Measure for Speech Recognition.
IEEE Trans. Acoustics, Speech and Signal Processing ASSP-35:1414-1422, Oct., 1987.

[83] A. Varga, R. Moore, J. Bridle, K. Ponting and M. Russell.
Noise Compensation Algorithms for Use with Hidden Markov Model Based Speech Recognition.
In *Proc. IEEE Int. Conf. Acoustics, Speech and Signal Processing, New York, NY*, pages 481-484. 1988.

[84] A. P. Varga and R. K. Moore.
Hidden Markov Model Decomposition of Speech and Noise.
In *Proc. IEEE Int. Conf. Acoustics, Speech and Signal Processing, Albuquerque, NM*, pages 845-848. April, 1990.

[85] W. Ward.
Modeling Non-Verbal Sounds for Speech Recognition.
In *Proc. Speech and Natural Language Workshop, Cape Cod, MA*, pages 311-318. Morgan Kaufmann, Oct., 1989.

[86] W. Ward.
Understanding Spontaneous Speech: The Phoenix System.
In *Proc. IEEE Int. Conf. Acoustics, Speech and Signal Processing, Toronto, Canada.* May, 1991.

[87] R. Zelinski.
A Microphone Array with Adaptive Post-Filtering for Noise Reduction in Reverberant Rooms.
In *Proc. IEEE Int. Conf. Acoustics, Speech and Signal Processing, New York, NY*, pages 2578-2581. 1988.

[88] V. Zue, J. Glass, D. Goodine, M. Phillips and S. Seneff.
The Summit Speech Recognition System: Phonological Modelling and Lexical Access.
In *Proc. IEEE Int. Conf. Acoustics, Speech and Signal Processing, Albuquerque, NM*, pages 49-52. April, 1990.

[89] E. Zwicker.
Subdivision of the Audible Frequency Range into Critical Bands.
Journal of the Acoustical Society of America (33):248-249, 1961.

Index